Glencoe
CHEMISTRY
MATTER AND CHANGE

Study Guide for Content Mastery

Student Edition

Glencoe
McGraw-Hill

New York, New York   Columbus, Ohio   Woodland Hills, California   Peoria, Illinois

A GLENCOE PROGRAM

Glencoe

CHEMISTRY
MATTER AND CHANGE

Hands-On Learning:
Laboratory Manual, SE/TE
Forensics Laboratory Manual, SE/TE
CBL Laboratory Manual, SE/TE
Small-Scale Laboratory Manual, SE/TE
ChemLab and MiniLab Worksheets

Review/Reinforcement:
Study Guide for Content Mastery, SE/TE
Solving Problems: A Chemistry Handbook
Reviewing Chemistry
Guided Reading Audio Program

Applications and Enrichment:
Challenge Problems
Supplemental Problems

Assessment:
Chapter Assessment
MindJogger Videoquizzes (VHS)
TestCheck Software, Windows/Macintosh

Teacher Resources:
Lesson Plans
Block Scheduling Lesson Plans
Spanish Resources
Section Focus Transparencies and Masters
Math Skills Transparencies and Masters
Teaching Transparencies and Masters
Solutions Manual

Technology:
Chemistry Interactive CD-ROM
Vocabulary PuzzleMaker Software,
Windows/Macintosh
Glencoe Science Web site:
science.glencoe.com

Send all inquiries to:
Glencoe/McGraw-Hill
8787 Orion Place
Columbus, OH 43240-4027

ISBN 0-07-824522-2
Printed in the United States of America.
12 13 14 045 10 09 08 07 06

Contents

To the Student

This *Study Guide for Content Mastery* for ***Chemistry: Matter and Change*** will help you learn more easily from your textbook. Each textbook chapter has six study guide pages of questions and exercises for you to complete as you read the text. The study guide pages are divided into sections that match those in your text.

You will find that the directions in the *Study Guide for Content Mastery* are simply stated and easy to follow. Sometimes you will be asked to answer questions. Other times, you will be asked to interpret a diagram or complete a table. By completing the study guide, you will gain a better understanding of the concepts presented in the text. These sheets also will prove helpful when studying for a test.

Before you begin your work, read the Study Skills section at the front of this booklet. The Study Skills section will help you

- improve your reading skills.
- improve your vocabulary skills.
- learn from visuals.
- make and understand idea maps.

These skills will help ensure your success in studying chemistry and any other discipline.

CREDITS

Art Credits
Navta Associates: **ix, 25, 35, 39, 71, 80, 85, 97, 114, 135;** Glencoe: **x, xi, 4, 10, 12, 24, 34, 53, 74, 75, 78, 98, 121, 125, 132, 154, 155, 156;** MacArt Design: **14, 48, 63, 92, 139, 143, 149**

Study Skills

A. Improve Your Reading Skills

Active readers are good readers.

Active readers

- get ready before they read.
- use skills that help them when they read.
- review to remember after they read.

Here's what you can do to become an active reader!

Before You Read

Get Ready to Read

- Find a quiet time and place to read—library, study hall, home.
- Don't read when you're tired.
- Don't read when you're hungry.
- Wait until you have finished a section before you take a break.

Scan

- Quickly scan the material so you will know what it is about.
- Look at pictures and read the captions, titles, headings, and words in bold print.

Write

- Write notes about what you see when you scan.
- Write questions about what you see.
- Write topics you want to find out about when you read.
- Write a preview outline from the section topics.

As You Read

- Find the main idea of each section or paragraph—this is usually in the first sentence.
- Study the pictures, maps, graphs, and tables, and think about the information in them.
- Write down the main ideas and other notes about what you read.
- After you read the whole section, reread the parts you didn't understand.

Study Skills

After You Read

- Review your outline or the notes you wrote while you were reading.
- If you still have questions, ask a classmate or your teacher for help.
- Write important facts or ideas on flash cards.
- Review your flash cards to help you remember what you've read.

B. Improve Your Vocabulary Skills

Active readers learn the meanings of new words.

Active readers

- recognize clues to help find the meaning.
- look for familiar words and word parts in new words.
- use a dictionary often.
- practice new words so they can remember new meanings.

Here's how you can improve your vocabulary!

When You See a New Word

Scan

- Read the sentence and look for clues about the meaning of the word. These are called context clues.
- Look for pictures or visuals that contain the word.

In the following table, you can find different kinds of context clues that you can use to help you figure out the meanings of new words.

Study Skills

Search for Context Clues

Comparison and contrast	The runner started the race with energy and excitement, but as she crossed the finish line, the *fatigue* and strain showed on her face.	This sentence contrasts the word *fatigue* with energy and compares it to strain. This tells you that someone who is fatigued is strained and has no energy.
Definition and description	Elena is a *chemist*, a scientist who studies matter and the changes that it undergoes.	The sentence describes a *chemist* as someone who studies matter and the changes that it undergoes.
Synonyms	Carl is very *dependable*. His teachers and his parents know that he is reliable and can be trusted.	The word *dependable* is described by the synonyms reliable and trusted.
Tone and setting	An air of *jubilation* surrounded the members of the science team as they received their medals for first place in the national competition.	The setting of the sentence and the action describe a situation that is positive and full of celebration.
A series of clues	The elements helium, oxygen, and nitrogen are all *nonmetals*.	The elements that are mentioned are all gases. This tells you something about the word *nonmetals*.
Cause and effect	The student group was known for its *boisterous* meetings, so the principal asked extra teachers to monitor the meeting and keep order.	*Boisterous* describes the meetings and tells you that something needs extra supervision.

Study Skills

Break It Down

- Find the root word.
- Write it and ask questions about its meaning.
- Find the affix—the part in front of or after the root word.
- Write it down and use a dictionary to look up its meaning.

public•ize

In this table, you can see how to break words into their roots and affixes.

Word	Root	Affix and Meaning	Meaning
imperfect	perfect	im- (not)	not perfect
semicircle	circle	semi- (half)	half of a circle
teacher	teach	-er (one who)	one who teaches
backward	back	-ward (in the direction of)	to the back
publicize	public	-ize (make)	make public

Remember New Words

- Say the word aloud.
- Write another sentence using the word.
- Make flash cards that include the word and its meaning.
- Review your flash cards to help you remember the meanings of the new words.

Study Skills

C. Learn From Visuals

Tables, graphs, photographs, diagrams, and maps are called visuals. Good readers use all kinds of visuals to help them learn.

Active readers

- find the purpose for the visual they see.
- find information in the visual.
- connect the information they find to what they are studying.

Here's how you can improve your skill in learning from visuals.

When You First Look at a Visual

Scan

- Look at the visual.
- Decide its purpose. Why is it there?
- Find the title.
- Read the caption.

Write

- Write the purpose of the visual. Why is it there?
- Write the key information.
- Write the title of the visual.
- Write the main idea or message.

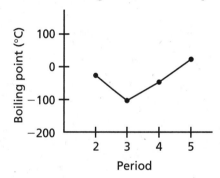

Boiling Point Periodicity

Study Skills

As You Study the Visual

Graphs

Graphs are pictures of related information. A graph tells you something about a specific situation. There are many kinds of graphs. One of the most common is the bar graph.

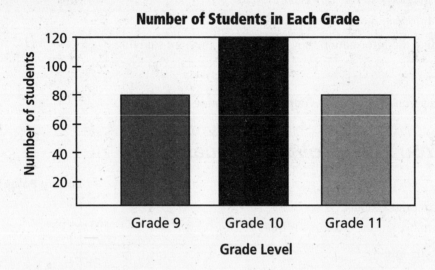

A bar graph helps you compare similar information about different items. The separate items being measured are shown as rectangles side by side on the graph.

Diagrams

A diagram is a drawing that has labels on it. It can show how something works or what the parts are called.

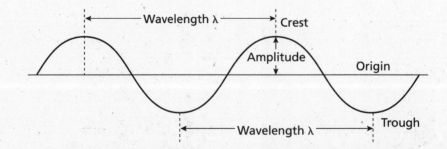

A diagram often gives the names of the parts of something, like this diagram of an electromagnetic wave. Science books often have many diagrams.

Study Skills

Tables

Tables organize words and numbers for easier reading. They have a title, columns (up and down), and rows (side to side). In this table, the columns show the innings, and the rows show the points each team scored.

Points Earned in the Baseball Game										
Inning	1	2	3	4	5	6	7	8	9	Total Points
Green Team	0	0	1	1	0	0	0	3	0	5
Blue Team	1	0	1	0	2	0	1	0	1	6

Maps

Maps give all kinds of different information. Some examples are location, direction, and land features. They can have words, symbols, numbers, lines, and colors.

Coal Fields of the United States

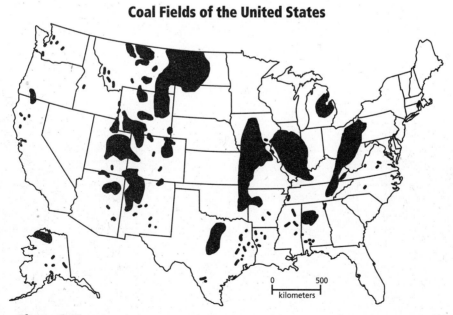

Figure 6.11

Coal is the most abundant fossil fuel on Earth. The coal deposits of the United States are mainly bituminous coal, which is preferred for electric power generation.

Study Skills

D. Make Chapter and Section Idea Maps

Active readers organize the information they read.

Active readers

- divide the information into smaller units.
- put the information in a logical order.

Starting Out

Scan and Write

- Scan the chapter for main topics and subheadings—in your chemistry textbook, blue headings are main topics and red headings are subtopics.
- Scan for boldface key terms.
- Scan for any visuals.
- Write the information in some kind of graphic map.

Here's an example of one kind of concept map.

Concept Map

```
        Blue                              Blue
       heading                           heading

  Red     Red     Red              Red     Red     Red
heading heading heading          heading heading heading

        Key                              Key
       terms                            terms
```

CHAPTER 1 STUDY GUIDE FOR CONTENT MASTERY

Introduction to Chemistry

Section 1.1 The Stories of Two Chemicals

In your textbook, read about the ozone layer.

Use each of the terms below just once to complete the passage.

atmosphere	oxygen gas	ozone	ozone hole
stratosphere	troposphere	ultraviolet radiation	

Earth's **(1)**_____ is made up of several layers. The air we breathe makes up the lowest level. This layer is called the **(2)**_____. The next layer up is called the **(3)**_____. This level contains a protective **(4)**_____ layer.

Ozone forms when **(5)**_____ is struck by ultraviolet radiation in the upper part of the stratosphere. The ozone forms a layer around Earth, which absorbs **(6)**_____. Without ozone, you are more likely to get a sunburn or possibly skin cancer. The thinning of the ozone layer, called the **(7)**_____, is worrisome because without ozone all organisms on Earth are subject to harm from too much radiation.

In your textbook, read about chlorofluorocarbons.

For each statement below, write *true* or *false*.

_____ **8.** CFC is another name for a chlorofluorocarbon.

_____ **9.** CFCs are made up of carbon, fluorine, and cesium.

_____ **10.** All CFCs are synthetic chemicals.

_____ **11.** CFCs usually react readily with other chemicals.

_____ **12.** CFCs were developed as replacements for toxic refrigerants.

CHAPTER 1   **STUDY GUIDE FOR CONTENT MASTERY**

Section 1.2 **Chemistry and Matter**

In your textbook, read about chemistry and matter.

Define each term.

1. chemistry

_____

2. matter

_____

3. mass

_____

Write each term below under the correct heading. Use each term only once.

air	magnetic field	car	feeling	heat	human body
light	radio	radio wave	flashlight	textbook	thought

Made of Matter	**Not Made of Matter**
4. _____	**10.** _____
5. _____	**11.** _____
6. _____	**12.** _____
7. _____	**13.** _____
8. _____	**14.** _____
9. _____	**15.** _____

For each statement below, write *true* or *false*.

_____ **16.** The mass of an object can vary with the object's location.

_____ **17.** A mass measurement includes the effect of Earth's gravitational pull on the object being measured.

_____ **18.** Scientists measure the amount of matter in terms of mass.

_____ **19.** Subtle differences in weight exist at different locations on Earth.

_____ **20.** Your mass on the Moon would be smaller than your mass on Earth.

Section 1.2 *continued*

Identify each branch of chemistry described.

21. The study of the matter and processes of living things

_____

22. The study of carbon-containing chemicals

_____

23. The study of the components and composition of substances

_____

24. The study of matter that does not contain organic chemicals

_____

25. The study of the behavior and changes of matter and the related energy changes

_____

For each branch of chemistry in Column A, write the letter of the item in Column B that pertains to that branch.

Column A	Column B
_____ **26.** Organic chemistry	**a.** reaction mechanisms
_____ **27.** Physical chemistry	**b.** minerals
_____ **28.** Biochemistry	**c.** plastics
_____ **29.** Analytical chemistry	**d.** metabolism
_____ **30.** Inorganic chemistry	**e.** quality control

Answer the following questions.

31. Compare the macroscopic world with the submicroscopic world.

_____

_____

_____

32. Why are chemists interested in the submicroscopic description of matter?

_____

_____

_____

CHAPTER 1 • **STUDY GUIDE FOR CONTENT MASTERY**

Section 1.3 **Scientific Methods**

In your textbook, read about a systematic approach that scientists use.

Use the words below to complete the concept map. Write your answers in the spaces below the concept map.

conclusions	experiments	hypothesis	scientific law	theory

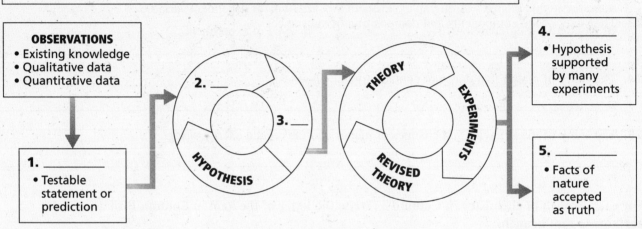

1. ___hypothesis___

2. ___experiments___

3. ___Conclusions___

4. ___theory___

5. ___scientific law___

For each item in Column A, write the letter of the matching item in Column B.

Column A

_b_ **6.** Refers to physical characteristics such as color, odor, or shape

_c_ **7.** Refers to mass, volume, and temperature measurements

_d_ **8.** A variable controlled by the experimenter

_a_ **9.** The act of gathering information

_e_ **10.** Changes in value based on the value of the controlled variable

Column B

a. observation

b. qualitative data

c. quantitative data

d. independent variable

e. dependent variable

CHAPTER 1

STUDY GUIDE FOR CONTENT MASTERY

Section 1.3 *continued*

Circle the letter of the choice that best completes the statement.

11. A constant is a factor that
  a. changes during an experiment.
  b. changes from one lab group to another.
  c. is affected by the dependent variable.
  d. is not allowed to change during an experiment.

12. A control is a
  a. variable that changes during an experiment.
  b. standard for comparison.
  c. type of dependent variable.
  d. type of experiment.

13. A hypothesis is a(n)
  a. set of controlled observations.
  b. explanation supported by many experiments.
  c. tentative explanation of observations.
  d. law describing a relationship in nature.

14. A theory is a(n)
  a. set of controlled observations.
  b. explanation supported by many experiments.
  c. tentative explanation of observations.
  d. law describing a relationship in nature.

15. A model is a(n)
  a. visual, verbal, and/or mathematical explanation of how things occur.
  b. explanation that is supported by many experiments.
  c. description of a relationship in nature.
  d. tentative explanation about what has been observed.

In the space at the left, write the word or phrase in parentheses that correctly completes the statement.

scientic method **16.** Molina and Rowland used a (model, scientific method) to learn about CFCs in the atmosphere.

Ultraviolent light ozone **17.** Their hypothesis was that CFCs break down in the stratosphere due to interactions with (ultraviolet light, oxygen).

18. Molina and Rowland thought that these interactions produced a chemical that could break down (chlorine, ozone).

hypothesis **19.** To test their (data, hypothesis), Molina and Rowland examined interactions that occur in the stratosphere.

model **20.** Based on their data, Molina and Rowland developed a (hypothesis, model) that explained how CFCs destroy ozone.

Chlorine **21.** Molina and Rowland concluded that (chlorine, radiation) formed by the breakdown of CFCs in the stratosphere reacts with ozone and destroys it.

Section 1.4 Scientific Research

In your textbook, read about types of scientific investigations.

For each description below, write *A* for applied research or *P* for pure research.

_____ **1.** Is undertaken to solve a specific problem

_____ **2.** Seeks to gain knowledge for the sake of knowledge itself

_____ **3.** Is used to find CFC replacements

_____ **4.** Was conducted by Molina and Rowland

In your textbook, read about students in the laboratory and the benefits of chemistry.

Answer the following questions.

5. When should you read the label on a chemical container?

_____

_____

6. What do scientists usually do when a scientific problem first arises?

_____

_____

7. What kinds of clothing should not be worn in the lab?

_____

8. What is technology?

_____

_____

9. Which type of research would you be more interested in working in—pure research or applied research? Why?

_____

_____

_____

CHAPTER **2** STUDY GUIDE FOR CONTENT MASTERY

Data Analysis

Section 2.1 Units of Measurement

In your textbook, read about SI units.

Complete the following table.

SI Base Units		
Quantity	**Base unit**	**Unit abbreviation**
1.		s
2. Mass		
3.	kelvin	
4. Length		

In your textbook, read about base units and derived units.

For each SI unit in Column A, write the letter of the matching item from Column B.

Column A	Column B
_____ **5.** second	**a.** A platinum-iridium cylinder that is stored at constant temperature and humidity
_____ **6.** meter	**b.** The microwave frequency given off by a cesium-133 atom
_____ **7.** kilogram	**c.** A cube whose sides all measure exactly one meter
_____ **8.** cubic meter	**d.** The distance that light travels through a vacuum in 1/299 792 458 second

9. Use Table 2–2 in your textbook to arrange the following prefixes in order from largest to smallest.

centi-   giga-   kilo-   mega-   milli-   nano-   pico-

_____

10. List the symbols and factors that the following prefixes represent.

   a. centi- _____

   b. kilo- _____

   c. milli- _____

Section 2.1 *continued*

Answer the following questions.

11. Which temperature scale will you use for your experiments in this class? Is this an SI unit?

_____

12. How many grams are in a kilogram?

_____

13. How many liters are in a megaliter?

_____

14. How many centimeters are in a meter?

_____

15. What is the difference between a base unit and a derived unit?

_____

_____

16. What is density?

_____

17. Explain in terms of density why a grocery bag containing all canned goods is harder to lift than a grocery bag containing all paper goods.

_____

_____

_____

18. How can you obtain an object's volume if you know its density and its mass?

_____

_____

_____

19. What is the three-part process for problem solving?

_____

20. How are degrees Celsius converted to kelvins?

_____

CHAPTER 2 **STUDY GUIDE FOR CONTENT MASTERY**

Section 2.2 Scientific Notation and Dimensional Analysis

In your textbook, read about scientific notation.

1. Circle the figures that are written in scientific notation.

1.61×10^2	$1.61 \times 10 \times 10$
1.61×100	161 km
$1.627\ 62 \times 10^{-27}$ kg	$9.109\ 39 \times 10^{-31}$ kg
2.8×10^{-8}	1 380 000

2. Change the following data into scientific notation.

a. 5 000 000 km _____

c. 0.000 421 g _____

b. 8 394 000 000 s _____

d. 0.03 cm _____

In your textbook, read about dimensional analysis.

Answer the following questions.

3. What is a conversion factor?

_____

_____

4. What is dimensional analysis?

_____

Complete the following dimensional analysis problems.

5. Convert 50 kilograms into grams.

50 _____ × 1000 _____/1 _____ = 50 000 _____

6. Convert 5 meters into centimeters.

5 _____ × 100 _____/1 _____ = 500 _____

7. Convert 5 liters into kiloliters.

5 _____ × 1 _____/1000 _____ = 0.0005 _____

8. Convert 5 centimeters into meters.

5 _____ × 1 _____/100 _____ = 0.05 _____

9. Convert 55 kilometers per hour into meters per second. Use the conversion factor 1 km = 1000 m.

55 _____/_____ × 1000 _____/1 _____ × 1 _____/60 _____

× 1 _____/60 _____ = 15 _____

Section 2.3 How reliable are measurements?

In your textbook, read about accuracy and precision.

1. Use the terms *precise* and *accurate* to describe the following figures. You may use both terms for some figures. If a term does not apply to a figure, leave the space blank.

a. _____   **b.** _____   **c.** _____

_____   _____   _____

Circle the letter of the choice that best completes the statement or answers the question.

2. The difference between an accepted value and an experimental value is called a(n)

  a. error.   **c.** measured value.

  b. percent error.   **d.** precise measurement.

3. The ratio of an error to an accepted value is called a(n)

  a. accuracy-to-precision value.   **c.** percent error.

  b. accuracy.   **d.** precision.

4. When you calculate percent error, you can ignore the

  a. accepted values.   **c.** experimental values.

  b. measured values.   **d.** plus and minus signs.

5. If two measurements are very close to each other, then they are

  a. accurate.   **c.** both accurate and precise.

  b. precise.   **d.** accepted values.

6. Which of the following is most likely to produce data that are not precise?

  a. a balance that is not set to zero

  b. not reading a graduated cylinder at eye level

  c. altering the procedure during an experiment

  d. making the same error with each trial

Section 2.3 *continued*

In your textbook, read about significant figures.

Use each of the terms below just once to complete the statements.

counting numbers	estimated	non-zero	zeros
scientific notation	significant figures	placeholders	

7. The digits that are reported in an answer are called _____.

8. The numeral 9.66 has three significant figures, two known figures and one

   _____ figure.

9. _____ numbers are always significant.

10. All final _____ to the right of the decimal place are significant.

11. Zeros that act as _____ are not significant.

12. _____ have an infinite number of significant figures.

13. When you convert to _____, you remove the placeholder zeros.

In your textbook, read about rounding off numbers.

14. Round the following to four significant figures.

   a. 12.555 km _____   **b.** 1.0009 _____   **c.** 99.999 _____   **d.** 23.342 999 _____

15. Round 12.783 456 to the requested number of significant figures.

   a. 2 significant figures _____   **c.** 6 significant figures _____

   b. 5 significant figures _____   **d.** 7 significant figures _____

16. Round 120.752 416 to the requested number of significant figures.

   a. 3 significant figures _____   **c.** 5 significant figures _____

   b. 4 significant figures _____   **d.** 7 significant figures _____

17. Complete the following calculations. Round off the answers to the correct number of significant figures.

   a. 51.2 kg + 64.44 kg _____

   b. 6.435 cm − 2.18 cm _____

   c. 16 m × 2.82 m × 0.05 m _____

   d. 3.46 m/1.82 s _____

CHAPTER 2   **STUDY GUIDE FOR CONTENT MASTERY**

Section 2.4 **Representing Data**

In your textbook, read about graphing.

Label each kind of graph shown.

1.   **Sources of Chlorine in the Stratosphere**

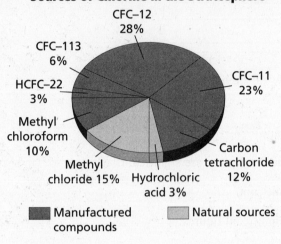

2.   **Precipitation in Jacksonville (1961–1990)**

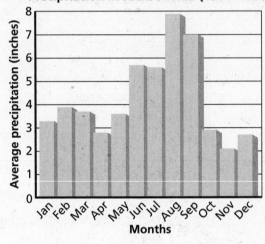

_____   _____

Answer the following questions about the graphs.

3. What percent of the sources of chlorine in the stratosphere are CFCs? _____

4. During which month of the year does Jacksonville usually get the most precipitation?
The least?

_____

In your textbook, read about line graphs.

**Sequence the following steps. Write *1* beside the first step in plotting a line graph.
Write *2* beside the second step, and so on.**

_____   **5.** Give the graph a title.

_____   **6.** Choose the ranges for the axes.

_____   **7.** Identify the independent and dependent variables.

_____   **8.** Plot the data points.

_____   **9.** Determine the range of the data that needs to be plotted
for each axis.

_____   **10.** Draw the "best fit" line for the data.

_____   **11.** Number and label each axis.

Name _____ Date _____ Class _____

Matter—Properties and Changes

Section 3.1 Properties of Matter

In your textbook, read about physical properties and chemical properties of matter.

Use each of the terms below just once to complete the passage.

chemical	mass	physical
density	properties	substance

Matter is anything with **(1)**_____ and volume. A

(2)_____ is a form of matter with a uniform and unchanging composition.

Substances have specific, unchanging **(3)**_____ that can be observed.

Substances have both physical and chemical properties. **(4)**_____

properties can be observed without changing a substance's chemical composition. Color,

hardness, and **(5)**_____ are examples. Other properties cannot be

observed without changing the composition of a substance. These are called

(6)_____ properties. An example is the tendency of iron to form

rust when exposed to air.

Label each property as either *physical* or *chemical*.

_____ **7.** Chemical formula H_2O

_____ **8.** Forms green carbonate when exposed to moist air

_____ **9.** Remains unchanged when in the presence of nitrogen

_____ **10.** Colorless

_____ **11.** Solid at normal temperatures and pressures

_____ **12.** Ability to combine with another substance

_____ **13.** Melting point

_____ **14.** Liquid at normal temperatures and pressures

_____ **15.** Boiling point is 100°C

_____ **16.** Conducts electricity

_____ **17.** Density is $\dfrac{1g}{cm^3}$

CHAPTER 3 STUDY GUIDE FOR CONTENT MASTERY

Section 3.1 *continued*

In your textbook, read about states of matter.

Label each drawing with one of these words: *solid, liquid, gas.*

18.

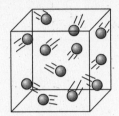

_____

19.

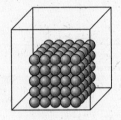

_____

20.

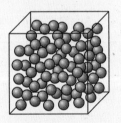

_____

For each statement below, write *true* **or** *false*.

_____ **21.** All matter that we encounter in everyday life exists in one of three physical forms.

_____ **22.** A solid has definite shape and volume.

_____ **23.** A liquid has a definite shape and takes on the volume of its container.

_____ **24.** A gas has both the shape and the volume of its container.

_____ **25.** The particles in a gas cannot be compressed into a smaller volume.

_____ **26.** Liquids tend to contract when heated.

_____ **27.** The particles in a solid are spaced far apart.

_____ **28.** The words *gas* and *vapor* can be used interchangeably.

Section 3.2 Changes in Matter

In your textbook, read about physical change and chemical change.

What kinds of changes do these words indicate? Write each word under the correct heading. Use each word only once.

boil	crumple	crush	explode
burn	ferment	freeze	grind
condense	melt	oxidize	rot
corrode	rust	tarnish	vaporize

Physical Change

1. _____

2. _____

3. _____

4. _____

5. _____

6. _____

7. _____

8. _____

Chemical Change

9. _____

10. _____

11. _____

12. _____

13. _____

14. _____

15. _____

16. _____

For each item in Column A, write the letter of the matching item in Column B.

Column A

_____ **17.** The new substances that are formed in a chemical reaction

_____ **18.** A chemical reaction that involves one or more substances changing into new substances

_____ **19.** Shows the relationship between the reactants and products in a chemical reaction

_____ **20.** States that mass is neither created nor destroyed in any process

_____ **21.** The starting substances in a chemical reaction

Column B

a. chemical change

b. reactants

c. products

d. chemical equation

e. law of conservation of mass

Answer the following question. Write an equation showing conservation of mass of reactants and products.

22. In a laboratory, 178.8 g of water is separated into hydrogen gas and oxygen gas. The hydrogen gas has a mass of 20.0 g. What is the mass of the oxygen gas produced?

_____

_____

CHAPTER 3 **STUDY GUIDE FOR CONTENT MASTERY**

Section 3.3 **Mixtures of Matter**

In your textbook, read about pure substances and mixtures.

Use the words below to complete the concept map.

heterogeneous	salt–water mixture	sand–water mixture
mixtures	solutions	water

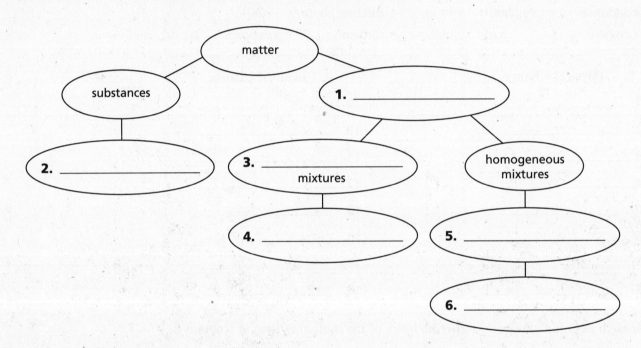

In your textbook, read about separating mixtures.

For each item in Column A, write the letter of the matching item in Column B.

Column A

_____ **7.** Separates substances on the basis of the boiling points of the substances

_____ **8.** Separates by formation of solid, pure particles from a solution

_____ **9.** Separates substances based on their movement through a special paper

_____ **10.** Separates solids from liquids by using a porous barrier

Column B

a. filtration

b. distillation

c. crystallization

d. chromatography

Section 3.4 Elements and Compounds

In your textbook, read about elements and compounds.

Circle the letter of the choice that best completes the statement or answers the question.

1. A substance that cannot be separated into simpler substances by physical or chemical means is a(n)

   a. compound. **b.** mixture. **c.** element. **d.** period.

2. A chemical combination of two or more different elements is a(n)

   a. solution. **b.** compound. **c.** element. **d.** period.

3. Which of the following is an example of an element?

   a. water **b.** air **c.** sugar **d.** oxygen

4. Which of the following is an example of a compound?

   a. gold **b.** silver **c.** aspirin **d.** copper

5. What are the horizontal rows in the periodic table called?

   a. block elements **b.** groups or families **c.** grids **d.** periods

6. What are the vertical columns in the periodic table called?

   a. block elements **b.** groups or families **c.** grids **d.** periods

Label each substance as either an *element* or a *compound*.

_____ **7.** silicon   _____ **10.** nickel

_____ **8.** sodium chloride   _____ **11.** ice

_____ **9.** francium

Write the symbol for each element. Use the periodic table on pages 72–73 in your textbook if you need help.

_____ **12.** neon   _____ **15.** titanium

_____ **13.** calcium   _____ **16.** fluorine

_____ **14.** iron

In your textbook, read about the law of definite proportions.

Use the law of definite proportions and the equation below to answer the questions.

The law of definite proportions states that regardless of the amount, a compound is always composed of the same elements in the same proportion by mass.

$$\text{Mass percentage of an element (\%)} = \frac{\text{mass of element}}{\text{mass of compound}} \times 100\%$$

17. A 20.0-g sample of sucrose contains 8.4 g of carbon. What is the mass percentage of carbon in sucrose? Show your work.

_____

STUDY GUIDE FOR CONTENT MASTERY

Section 3.4 *continued*

18. Sucrose is 51.50% oxygen. How many grams of oxygen are in 20.0 g of sucrose? Show your work.

_____

_____

19. A 2-g sample of sucrose is 6.50% hydrogen. What is the mass percentage of hydrogen in 300 g of sucrose? Explain your reasoning.

_____

_____

20. Two compound samples are found to have the same mass percentages of the same elements. What can you conclude about the two samples?

_____

In your textbook, read about the law of multiple proportions.

Use the law of multiple proportions to answer the questions and complete the table below.

The law of multiple proportions states that if the elements X and Y form two compounds, the different masses of Y that combine with a fixed mass of X can be expressed as a ratio of small whole numbers.

21. Two compound samples are composed of the same elements, but in different proportions. What can you conclude about the two samples?

_____

For each compound in the table, fill in the ratio of the mass of oxygen to the mass of hydrogen.

Compound	Mass of Oxygen	Mass of Hydrogen	Mass O/Mass H
H_2O	16 g	2 g	**22.**
H_2O_2	32 g	2 g	**23.**

24. Write a brief statement comparing the two mass ratios from the table.

_____

_____

25. Are H_2O and H_2O_2 the same compound? Explain your answer.

_____

_____

The Structure of the Atom

Section 4.1 Early Theories of Matter

In your textbook, read about the philosophers, John Dalton, and defining the atom.

For each statement below, write *true* or *false*.

_____ **1.** Ancient philosophers regularly performed controlled experiments.

_____ **2.** Philosophers formulated explanations about the nature of matter based on their own experiences.

_____ **3.** Both Democritus and Dalton suggested that matter is made up of atoms.

_____ **4.** Dalton's atomic theory stated that atoms separate, combine, or rearrange in chemical reactions.

_____ **5.** Dalton's atomic theory stated that matter is mostly empty space.

_____ **6.** Dalton was correct in thinking that atoms could not be divided into smaller particles.

_____ **7.** Dalton's atomic theory stated that atoms of different elements combine in simple whole-number ratios to form compounds.

_____ **8.** Dalton thought that all atoms of a specific element have the same mass.

_____ **9.** Democritus proposed that atoms are held together by chemical bonds, but no one believed him.

_____ **10.** Dalton's atomic theory was based on careful measurements and extensive research.

_____ **11.** There are no instruments powerful enough to magnify atoms so that they can be seen.

_____ **12.** The smallest particle of an element that retains the properties of that element is called an atom.

Section 4.2 Subatomic Particles and the Nuclear Atom

In your textbook, read about discovering the electron and the nuclear atom.

For each item in Column A, write the letter of the matching item in Column B.

Column A	Column B
_____ **1.** Proposed the nuclear atomic model	**a.** Thomson
_____ **2.** Determined the mass-to-charge ratio of an electron	**b.** Millikan
_____ **3.** Calculated the mass of an electron	**c.** Rutherford

Draw and label a diagram of each atomic model.

  4. plum pudding model

  5. nuclear atomic model

In your textbook, read about the discovery of protons and neutrons.

Complete the following table of proton, electron, and neutron characteristics.

Particle	Symbol	Location	Relative Charge	Relative Mass
6. Proton				
7.	n^0			
8.				1/1840

CHAPTER 4   **STUDY GUIDE FOR CONTENT MASTERY**

Section 4.3 How Atoms Differ

In your textbook, read about atomic number.

For each statement below, write *true* **or** *false.*

_____ **1.** The number of neutrons in an atom is referred to as its atomic number.

_____ **2.** The periodic table is arranged by increasing atomic number.

_____ **3.** Atomic number is equal to the number of electrons in an atom.

_____ **4.** The number of protons in an atom identifies it as an atom of a particular element.

_____ **5.** Most atoms have either a positive or a negative charge.

Answer the following questions.

6. Lead has an atomic number of 82. How many protons and electrons does lead have?

_____

7. Oxygen has 8 electrons. How many protons does oxygen have? _____

8. Zinc has 30 protons. What is its atomic number? _____

9. Astatine has 85 protons. What is its atomic number? _____

10. Rutherfordium has an atomic number of 104. How many protons and electrons does it have?

_____

11. Polonium has an atomic number of 84. How many protons and electrons does it have?

_____

12. Nobelium has an atomic number of 102. How many protons and electrons does it have?

_____

In your textbook, read about isotopes and mass number.

Determine the number of protons, electrons, and neutrons for each isotope described below.

13. An isotope has atomic number 19 and mass number 39.

_____

14. An isotope has 14 electrons and a mass number of 28.

_____

15. An isotope has 21 neutrons and a mass number of 40.

_____

Section 4.3 *continued*

16. An isotope has an atomic number 51 and a mass number 123.

_____

Answer the following question.

17. Which of the isotopes in problems 13–16 are isotopes of the same element? Identify the element.

_____

Write each isotope below in symbolic notation. Use the periodic table to determine the atomic number of each isotope.

18. neon-22 _____   **20.** cesium-133 _____

19. helium _____   **21.** uranium-234 _____

Label the mass number and the atomic number on the following isotope notation.

22. _____ ⟶ $^{24}_{12}Mg$

23. _____ ⟶

In your textbook, read about mass of individual atoms.

Circle the letter of the choice that best completes the statement.

24. The mass of an electron is

    a. smaller than the mass of a proton.   **c.** a tiny fraction of the mass of an atom.

    b. smaller than the mass of a neutron.   **d.** all of the above.

25. One atomic mass unit is

    a. 1/12 the mass of a carbon-12 atom.

    b. 1/16 the mass of an oxygen-16 atom.

    c. exactly the mass of one proton.

    d. approximately the mass of one proton plus one neutron.

26. The atomic mass of an atom is usually not a whole number because it accounts for

    a. only the relative abundance of the atom's isotopes.

    b. only the mass of each of the atom's isotopes.

    c. the mass of the atom's electrons.

    d. both the relative abundance and the mass of each of the atom's isotopes.

Section 4.3 *continued*

Use the figures to answer the following questions.

Osmium
76
Os
190.2

Niobium
41
Nb
92.906

27. What is the atomic number of osmium? _____

28. What is the chemical symbol for niobium? _____

29. What is the atomic mass of osmium? _____

30. What units is the atomic mass reported in? _____

31. How many protons and electrons does an osmium atom have? A niobium atom?

_____

Calculate the atomic mass of each element described below. Then use the periodic table to identify each element.

32.

Isotope	Mass (amu)	Percent Abundance
^{63}X	62.930	69.17
^{65}X	64.928	30.83

33.

Isotope	Mass (amu)	Percent Abundance
^{35}X	34.969	75.77
^{37}X	36.966	24.23

CHAPTER 4 **STUDY GUIDE FOR CONTENT MASTERY**

Section 4.4 Unstable Nuclei and Radioactive Decay

In your textbook, read about radioactivity.

For each item in Column A, write the letter of the matching item in Column B.

Column A		Column B
_____ **1.** The rays and particles that are emitted by a radioactive material		**a.** nuclear reaction
_____ **2.** A reaction that involves a change in an atom's nucleus		**b.** beta radiation
_____ **3.** The process in which an unstable nucleus loses energy spontaneously		**c.** radiation
_____ **4.** Fast-moving electrons		**d.** radioactive decay

In your textbook, read about types of radiation.

Use the diagram to answer the questions.

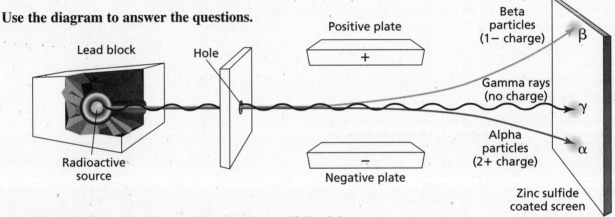

5. Which plate do the beta particles bend toward? Explain.

_____

6. Explain why the gamma rays do not bend.

_____

7. Explain why the path of the beta particles bends more than the path of the alpha particles.

_____

_____

Complete the following table of the characteristics of alpha, beta, and gamma radiation.

Radiation Type	Composition	Symbol	Mass (amu)	Charge
8. Alpha				
9.			1/1840	
10.	High-energy electromagnetic radiation			

Name _____ Date _____ Class _____

Electrons in Atoms

Section 5.1 Light and Quantized Energy

In your textbook, read about the wave nature of light.

Use each of the terms below just once to complete the passage.

amplitude	energy	frequency	hertz
light	wave	wavelength	speed

Electromagnetic radiation is a kind of **(1)**_____ that behaves like a(n)

(2)_____ as it travels through space. **(3)**_____ is one type of

electromagnetic radiation. Other examples include X rays, radio waves, and microwaves.

All waves can be characterized by their wavelength, amplitude, frequency, and

(4)_____. The shortest distance between equivalent points on a continuous wave is

called a(n) **(5)**_____. The height of a wave from the origin to a crest or from the

origin to a trough is the **(6)**_____. **(7)**_____ is the number of

waves that pass a given point in one second. The SI unit for frequency is the

(8)_____, which is equivalent to one wave per second.

Use the figure to answer the following questions.

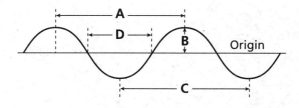

9. Which letter(s) represent one wavelength? _____

10. Which letter(s) represent the amplitude? _____

11. If twice the length of A passes a stationary point every second, what is the frequency of
the wave?

_____

Section 5.1 *continued*

In your textbook, read about the particle nature of light.

Circle the letter of the choice that best completes the statement or answers the question.

12. A(n) _____ is the minimum amount of energy that can be lost or gained by an atom.

    a. valence electron   **b.** electron   **c.** quantum   **d.** Planck's constant

13. According to Planck's theory, for a given frequency, v, matter can emit or absorb energy only in

    a. units of hertz.   **c.** entire wavelengths.

    b. whole-number multiples of hv.   **d.** multiples of $\frac{1}{2}hv$, $\frac{1}{4}hv$, and so on.

14. The _____ is the phenomenon in which electrons are emitted from a metal's surface when light of a certain frequency shines on it.

    a. quantum   **b.** Planck concept   **c.** photon effect   **d.** photoelectric effect

15. Which equation would you use to calculate the energy of a photon?

    a. $E_{photon} = hv \times$ Planck's constant   **c.** $E_{photon} = \frac{1}{2}hv$

    b. $E_{photon} = hv$ **d.** $c = \lambda v$

In your textbook, read about atomic emission spectra.

For each statement below, write *true* or *false*.

_____ **16.** Like the visible spectrum, an atomic emission spectrum is a continuous range of colors.

_____ **17.** Each element has a unique atomic emission spectrum.

_____ **18.** A flame test can be used to identify the presence of certain elements in a compound.

_____ **19.** The fact that only certain colors appear in an element's atomic emission spectrum indicates that only certain frequencies of light are emitted.

_____ **20.** Atomic emission spectra can be explained by the wave model of light.

_____ **21.** The neon atoms in a neon sign emit their characteristic color of light as they absorb energy.

_____ **22.** When an atom emits light, photons having certain specific energies are being emitted.

CHAPTER **5**   **STUDY GUIDE FOR CONTENT MASTERY**

Section 5.2 **Quantum Theory and the Atom**

In your textbook, read about the Bohr model of the atom.

Use each of the terms below to complete the statements.

atomic emission spectrum	electron	frequencies	ground state
higher	energy levels	lower	

1. The lowest allowable energy state of an atom is called its _____.

2. Bohr's model of the atom predicted the _____ of the lines in

   hydrogen's atomic emission spectrum.

3. According to Bohr's atomic model, the smaller an electron's orbit, the

   _____ the atom's energy level.

4. According to Bohr's atomic model, the larger an electron's orbit, the

   _____ the atom's energy level.

5. Bohr proposed that when energy is added to a hydrogen atom, its

   _____ moves to a higher-energy orbit.

6. According to Bohr's atomic model, the hydrogen atom emits a photon corresponding to

   the difference between the _____ associated with the two

   orbits it transitions between.

7. Bohr's atomic model failed to explain the _____ of elements

   other than hydrogen.

In your textbook, read about the quantum mechanical model of the atom.

Answer the following questions.

8. If you looked closely, could you see the wavelength of a fast-moving car? Explain
   your answer.

   _____

   _____

9. Using de Broglie's equation, $\lambda = \dfrac{h}{mv}$ which would have the larger wavelength, a

   slow-moving proton or a fast-moving golf ball? Explain your answer.

   _____

   _____

Section 5.2 *continued*

In your textbook, read about the Heisenberg uncertainty principle.

For each item in Column A, write the letter of the matching item in Column B.

Column A	Column B
_____ **10.** The modern model of the atom that treats electrons as waves	**a.** Heisenberg uncertainty principle
_____ **11.** States that it is impossible to know both the velocity and the position of a particle at the same time	**b.** Schrödinger wave equation
_____ **12.** A three-dimensional region around the nucleus representing the probability of finding an electron	**c.** quantum mechanical model of the atom
_____ **13.** Originally applied to the hydrogen atom, it led to the quantum mechanical model of the atom	**d.** atomic orbital

Answer the following question.

14. How do the Bohr model and the quantum mechanical model of the atom differ in how they describe electrons?

_____

_____

_____

In your textbook, read about hydrogen's atomic orbitals.

In the space at the left, write the term in parentheses that correctly completes the statement.

_____ **15.** Atomic orbitals (do, do not) have an exactly defined size.

_____ **16.** Each orbital may contain at most (two, four) electrons.

_____ **17.** All s orbitals are (spherically shaped, dumbbell shaped).

_____ **18.** A principal energy has (n, n^2) energy sublevels.

_____ **19.** The maximum number of (electrons, orbitals) related to each principal energy level equals $2n^2$.

_____ **20.** There are (three, five) equal energy p orbitals.

_____ **21.** Hydrogen's principal energy level 2 consists of (2s and 3s, 2s and 2p) orbitals.

_____ **22.** Hydrogen's principal energy level 3 consists of (nine, three) orbitals.

Section 5.3 **Electron Configurations**

In your textbook, read about ground-state electron configurations.

Use each of the terms below just once to complete the passage.

Aufbau principle	electron configuration	ground-state electron configuration	Hund's rule
lowest	Pauli exclusion principle	spins	stable

The arrangement of electrons in an atom is called the atom's

(1)_____. Electrons in an atom tend to assume the arrangement

that gives the atom the **(2)**_____ possible energy. This arrangement

of electrons is the most **(3)**_____ arrangement and is called the

atom's **(4)**_____.

Three rules define how electrons can be arranged in an atom's orbitals. The

(5)_____ states that each electron occupies the lowest energy

orbital available. The **(6)**_____ states that a maximum of two

electrons may occupy a single atomic orbital, but only if the electrons have opposite

(7)_____. **(8)**_____ states that single

electrons with the same spin must occupy each equal-energy orbital before additional

electrons with opposite spins occupy the same orbitals.

Complete the following table.

Element	Atomic Number	Orbitals					Electron Configuration
		1s	2s	2p$_x$	2p$_y$	2p$_z$	
9. Helium							1s^2
10.	7						
11. Neon		↑↓	↑↓	↑↓	↑↓	↑↓	

CHAPTER 5   **STUDY GUIDE FOR CONTENT MASTERY**

Section 5.3 *continued*

Answer the following questions.

12. What is germanium's atomic number? How many electrons does germanium have?

_____

13. What is noble-gas notation, and why is it used to write electron configurations?

_____

_____

_____

_____

14. Write the ground-state electron configuration of a germanium atom, using noble-gas notation.

_____

In your textbook, read about valence electrons.

Circle the letter of the choice that best completes the statement or answers the question.

15. The electrons in an atom's outermost orbitals are called

    a. electron dots.   **b.** quantum electrons.   **c.** valence electrons.   **d.** noble-gas electrons.

16. In an electron-dot structure, the element's symbol represents the

    a. nucleus of the noble gas closest to the atom in the periodic table.

    b. atom's nucleus and inner-level electrons.

    c. atom's valence electrons.

    d. electrons of the noble gas closest to the atom in the periodic table.

17. How many valence electrons does a chlorine atom have if its electron configuration is $[Ne]3s^23p^5$?

    a. 3   **b.** 21   **c.** 5   **d.** 7

18. Given boron's electron configuration of $[He]2s^22p^1$, which of the following represents its electron-dot structure?

    a. •Be•   **b.** •B̊•   **c.** B̈:   **d.** B̈e

19. Given beryllium's electron configuration of $1s^22s^2$, which of the following represents its electron-dot structure?

    a. •Be•   **b.** •B̊•   **c.** B̈:   **d.** B̈e

20. Which electrons are represented by the dots in an electron-dot structure?

    a. valence electrons   **c.** only s electrons

    b. inner-level electrons   **d.** both a and c

The Periodic Table and Periodic Law

Section 6.1 Development of the Modern Periodic Table

In your textbook, reads about the history of the periodic table's development.

Use each of the terms below just once to complete the passage.

octaves	atomic mass	atomic number	nine
elements	properties	Henry Moseley	eight
protons	periodic law	Dmitri Mendeleev	accepted

The table below was developed by John Newlands and is based on a relationship called

the law of **(1)**_____. According to this law, the properties of the elements

repeated every **(2)**_____ elements. Thus, for example, element two and

element **(3)**_____ have similar properties. The law of octaves did not work

for all the known elements and was not generally **(4)**_____.

1	2	3	4	5	6	7
H	Li	G	Bo	C	N	O
8	9	10	11	12	13	14
F	Na	Mg	Al	Si	P	S

The first periodic table is mostly credited to **(5)**_____. In his table, the

elements were arranged according to increasing **(6)**_____. One important

result of this table was that the existence and properties of undiscovered

(7)_____ could be predicted.

The element in the modern periodic table are arranged according to increasing

(8)_____, as a result of the work of **(9)**_____. This

arrangement is based on number of **(10)**_____ in the nucleus of an atom of

the element. The modern form of the periodic table results in the

(11)_____, which states that when elements are arranged according to

increasing atomic number, there is a periodic repetition of their chemical and physical

(12)_____.

Section 6.1 *continued*

In your textbook, read about the modern periodic table.

Use the information in the box on the left taken from the periodic table to complete the table on the right.

7
N
Nitrogen
14.007
[He]$2s^2 2p^3$

Atomic Mass	**13.**
Atomic Number	**14.**
Electron Configuration	**15.**
Chemical Name	**16.**
Chemical Symbol	**17.**

For each item in Column A, write the letter of the matching item in Column B.

Column A	Column B
_____ **18.** A column on the periodic table	**a.** metals
_____ **19.** A row on the periodic table	**b.** group
_____ **20.** Group A elements	**c.** period
_____ **21.** Elements that are shiny and conduct electricity	**d.** representative elements
_____ **22.** Group B elements	**e.** transition elements

In the space at the left, write *true* **if the statement is true; if the statement is false, change the italicized word or phrase to make it true.**

_____ **23.** There are *two* main classifications of elements.

_____ **24.** More than three-fourths of the elements in the periodic table are *nonmetals.*

_____ **25.** Group 1A elements (except for hydrogen) are known as the *alkali metals.*

_____ **26.** *Group 3A* elements are the alkaline earth metals.

_____ **27.** Group 7A elements are highly reactive nonmetals known as *halogens.*

_____ **28.** Group 8A elements are very unreactive elements known as *transition metals.*

_____ **29.** Metalloids have properties of both metals and *inner transition metals.*

CHAPTER 6 **STUDY GUIDE FOR CONTENT MASTERY**

Section 6.2 Classification of the Elements

In your textbook, read about organizing the elements by electron configuration.

Use the periodic table on pages 156–157 in your textbook to match each element in Column A with the element in Column B that has the most similar chemical properties.

	Column A		Column B
_____	**1.** arsenic (As)	**a.**	boron (B)
_____	**2.** bromine (Br)	**b.**	cesium (Cs)
_____	**3.** cadmium (Cd)	**c.**	chromium (Cr)
_____	**4.** gallium (Ga)	**d.**	cobalt (Co)
_____	**5.** germanium (Ge)	**e.**	hafnium (Hf)
_____	**6.** iridium (Ir)	**f.**	iodine (I)
_____	**7.** magnesium (Mg)	**g.**	iron (Fe)
_____	**8.** neon (Ne)	**h.**	nitrogen (N)
_____	**9.** nickel (Ni)	**i.**	platinum (Pt)
_____	**10.** osmium (Os)	**j.**	scandium (Sc)
_____	**11.** sodium (Na)	**k.**	silicon (Si)
_____	**12.** tellurium (Te)	**l.**	strontium (Sr)
_____	**13.** tungsten (W)	**m.**	sulfur (S)
_____	**14.** yttrium (Y)	**n.**	zinc (Z)
_____	**15.** zirconium (Zr)	**o.**	xenon (Xe)

Answer the following questions.

16. Why do sodium and potassium, which belong to the same group in the periodic table, have similar chemical properties?

_____

_____

17. How is the energy level of an element's valence electrons related to its period on the periodic table? Give an example.

_____

_____

Section 6.2 *continued*

In your textbook, read about s-, p-, d-, and f-block elements.

Use the periodic table on pages 156–157 in your textbook and the periodic table below to answer the following questions.

s block												p block						s² 2 He

(periodic table diagram showing s-block, p-block, d-block, and f-block elements)

18. Into how many blocks is the periodic table divided? _____

19. What groups of elements does the s-block contain? _____

20. Why does the s-block portion of the periodic table span two groups?

_____

21. What groups of elements does the p-block contain? _____

22. Why are members of group 8A virtually unreactive?

_____

_____

_____

23. How many d-block elements are there? _____

24. What groups of elements does the d-block contain? _____

25. Why does the f-block portion of the periodic table span 14 groups?

_____

26. What is the electron configuration of the element in period 3, group 6A? _____

Name _____ Date _____ Class _____

Section 6.3 **Periodic Trends**

In your textbook, read about atomic radius and ionic radius.

Circle the letter of the choice that best completes the statement or answers the question.

1. Atomic radii cannot be measured directly because the electron cloud surrounding the nucleus does not have a clearly defined

   a. charge.   **b.** mass.   **c.** outer edge.   **d.** probability.

2. Which diagram best represents the group and period trends in atomic radii in the periodic table?

   a.
   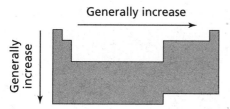

   c.
   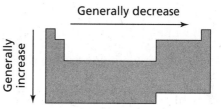

   b.
   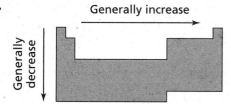

   d.
   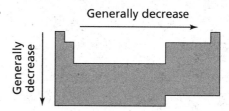

3. The general trend in the radius of an atom moving down a group is partially accounted

   for by the

   a. decrease in the mass of the nucleus.   **c.** increase in the charge of the nucleus.

   b. fewer number of filled orbitals.   **d.** shielding of the outer electrons by inner electrons.

4. A(n) _____ is an atom, or bonded group of atoms, that has a positive or negative charge.

   a. halogen   **b.** ion   **c.** isotope   **d.** molecule

5. An atom becomes negatively charged by

   a. gaining an electron.   **b.** gaining a proton.   **c.** losing an electron.   **d.** losing a neutron.

6. Which diagram best represents the relationship between the diameter of a sodium atom and the diameter of a positive sodium ion?

   a.   **b.** ◯ ◯   **c.** ◯ ◯
     Na   Na⁺   Na   Na⁺   Na   Na⁺

Section 6.3 *continued*

In your textbook, read about ionization energy and electronegativity.

Answer the following questions.

7. What is ionization energy?

_____

_____

8. Explain why an atom with a high ionization-energy value is not likely to form a positive ion.

_____

_____

9. What is the period trend in the first ionization energies? Why?

_____

_____

_____

10. What is the group trend in the first ionization energies? Why?

_____

_____

_____

_____

11. State the octet rule.

_____

_____

12. What does the electronegativity of an element indicate?

_____

_____

13. What are the period and group trends in electronegativities?

_____

_____

CHAPTER **7** **STUDY GUIDE FOR CONTENT MASTERY**

The Elements

Section 7.1 Properties of s-Block Elements

In your textbook, read about the representative elements and hydrogen.

Answer the following questions.

1. Why are the elements in groups 1A–8A called the representative elements?

_____

_____

_____

2. What do all the elements in a group have in common?

_____

3. How many valence electrons do group 1A elements have? Group 2A elements?

_____

_____

4. What is the difference between the electron configurations for elements in groups 1A and 2A and those for elements in groups 3A–8A?

_____

_____

5. Why don't all the elements in a group have the same properties?

_____

_____

6. Why is hydrogen not considered a part of any group?

_____

_____

7. How does hydrogen act like a metal? A nonmetal?

_____

_____

8. Name two ways hydrogen is produced in the lab.

_____

_____

Section 7.1 *continued*

In your textbook, read about the alkali and alkaline earth metals.

For each statement below, write *true* or *false*.

_____ **9.** Alkali and alkaline earth metals are not reactive.

_____ **10.** Alkali metals are shiny gray solids soft enough to be cut by a knife.

_____ **11.** The metal lithium has a diagonal relationship with the metal magnesium.

_____ **12.** An alloy is a compound of a metal and oxygen.

_____ **13.** Sodium and potassium are the most abundant alkali metals.

_____ **14.** Potassium chloride can be used as a substitute for sodium chloride.

_____ **15.** Alkaline earth metals form ions with a +2 charge.

_____ **16.** The order of reactivity of the alkaline earth metals, from most to least reactive, is beryllium, magnesium, calcium, strontium, and barium.

_____ **17.** Alkaline earth metals do not react with oxygen.

For each item in Column A, write the letter of the matching item in Column B.

Column A	Column B
_____ **18.** nonsparking tools	**a.** sodium
_____ **19.** fireworks	**b.** calcium
_____ **20.** fertilizers	**c.** barium
_____ **21.** chlorophyll	**d.** beryllium
_____ **22.** chalk	**e.** magnesium
_____ **23.** X rays	**f.** strontium
_____ **24.** treating bipolar disorders	**g.** potassium
_____ **25.** baking soda	**h.** lithium

Circle the element that would be the most reactive of the three.

26. Li   Na   K

27. K   Fr   Na

28. Ba   Mg   Be

29. Li   K   Ba

CHAPTER (7)   **STUDY GUIDE FOR CONTENT MASTERY**

Section 7.2 Properties of the p-Block Elements

In your textbook, read about the boron and carbon groups.

Write the word(s) or number that best completes the statement.

1. The p-block elements contain metals, _____, nonmetals, and

     _____.

2. Boron is found in California in the form of _____.

3. _____ is the most abundant metal in Earth's crust.

4. Boron nitride and aluminum oxide are both _____.

5. Graphite and diamond are examples of _____ because they are
different forms of the same element in the same state.

6. The branch of chemistry that studies most carbon compounds is _____.

7. Carbonates, cyanides, carbides, sulfides, and oxides of carbon are classified as inorganic

compounds. Geologists call these substances _____.

8. The second most abundant element in Earth's crust is _____.

9. Computer chips made of silicon are less efficient than _____ chips.

Use the structures below to answer the following questions.

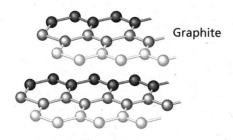
Graphite

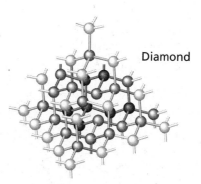
Diamond

10. What do these structures have in common? What is different about them?

     _____

     _____

11. Explain why graphite is a good lubricant and diamond is an excellent abrasive.

     _____

     _____

Section 7.2 *continued*

In your textbook, read about the nitrogen group, the oxygen group, and the halogens.

Circle the letter of the choice that best completes the statement or answers the question.

12. Plants and animals get the nitrogen they need
  a. directly from the air.
  b. from nitrogen-fixing bacteria.
  c. from ammonia compounds.
  d. nitric acid.

13. The major industrial use for nitrogen is to make
  a. ammonia.   **b.** nitrous oxide.   **c.** nitrous acid.   **d.** nitric oxide.

14. Some of the uses for phosphorous are
  a. in cleaning products and fertilizers.
  b. in lightweight metals and tools.
  c. in ceramics and food seasonings.
  d. in eyebrow pencils and storage batteries.

15. Which of the following is used as an energy source by certain bacteria living near ocean vents?
  a. oxygen   **b.** hydrogen sulfide   **c.** water   **d.** carbon dioxide

16. Oxygen is the most abundant element
  a. in the universe.
  b. in Earth's atmosphere.
  c. in Earth's crust.
  d. in living organisms.

17. Ozone is
  a. an isotope of oxygen.
  b. a compound of oxygen.
  c. an allotrope of oxygen.
  d. a stable gas.

18. The release of phosphate ions from fertilizers or detergents into bodies of water may lead to depletion of dissolved oxygen because
  a. bacteria in the water decompose phosphate ions.
  b. the phosphate ions form a layer in the water.
  c. phosphate ions serve as nutrients for algae.
  d. the phosphate ions replace oxygen dissolved in the water.

19. Selenium is used in solar panels because it
  a. can convert light into electricity.
  b. is a metal.
  c. is heavy.
  d. can convert light into heat.

20. The most chemically active of all the elements is
  a. sodium.   **b.** carbon.   **c.** fluorine.   **d.** bromine.

21. Some of the uses for chlorine are in
  a. metal alloys.
  b. fertilizers and photographic materials.
  c. toothpaste and cookware.
  d. bleaches, disinfectants, and plastics.

CHAPTER 7

STUDY GUIDE FOR CONTENT MASTERY

Section 7.3 Properties of d-Block and f-Block Elements

In your textbook, read about the transition and inner transition metals.

Answer the following questions.

1. What is the difference between the transition metals and the inner transition metals?

   _____

   _____

2. Write the electron configurations for titanium (Ti), vanadium (V), chromium (Cr), and manganese (Mn).

   _____

   _____

3. Which of the elements listed in question 2 would be the hardest? Which would have the highest melting and boiling points? Why?

   _____

   _____

4. Which of the elements would be least likely to form compounds with color? Why?

   _____

   _____

5. How are differences in properties among the transition elements explained?

   _____

6. How can a transition metal form an ion with a charge of 3+ or higher?

   _____

   _____

7. Identify three ways transition metals are separated from their ores.

   _____

8. Explain how a metal can become a temporary magnet.

   _____

   _____

   _____

9. Explain why some metals can act as permanent magnets.

   _____

   _____

Name _____ Date _____ Class _____

Section 7.3 *continued*

For each item in Column A, write the letter of the matching item in Column B. The items in Column B can be used more than once.

Column A	Column B
_____ **10.** Found in the center of a hemoglobin molecule	**a.** tungsten
_____ **11.** Provides a protective coating to resist rusting	**b.** chromium
_____ **12.** Used in electrical wiring	**c.** zinc
_____ **13.** Needed for the development of red blood cells	**d.** cobalt
_____ **14.** Can control the conditions at which a reaction occurs	**e.** iron
_____ **15.** Involved in cell respiration	**f.** platinum
_____ **16.** The main element in steel	**g.** manganese
_____ **17.** Are classified as "strategic" metals	**h.** copper
_____ **18.** Found in molecules that help the body digest proteins and eliminate carbon dioxide	**i.** silver
_____ **19.** The best conductor of electricity	

In your textbook, read about the inner transition metals.

Write the word or words that best complete the statement.

20. Because there is little difference in the properties of the _____, they are usually found mixed together in nature.

21. Welders' goggles contain _____ and _____ because these elements absorb high energy radiation.

22. Because some of its compounds emit red light when excited by electrons, _____ is often used in TV screens.

23. The _____ are radioactive.

24. Elements with atomic numbers above 92 are called _____.

25. Some smoke detectors used in the home contain _____.

CHAPTER 8 STUDY GUIDE FOR CONTENT MASTERY

Ionic Compounds

Section 8.1 Forming Chemical Bonds

In your textbook, read about chemical bonds and formation of ions.

Use each of the terms below just once to complete the passage.

chemical bond	electrons	energy level	ions	noble gases
nucleus	octet	pseudo-noble gas formations		valence

The force that holds two atoms together is called a(n) **(1)** _____.

Such an attachment may form by the attraction of the positively charged

(2) _____ of one atom for the negatively charged

(3) _____ of another atom, or by the attraction of charged atoms,

which are called **(4)** _____. The attractions may also involve

(5) _____ electrons, which are the electrons in the outermost

(6) _____. The **(7)** _____ are a family of elements that

have very little tendency to react. Most of these elements have a set of eight outermost

electrons, which is called a stable **(8)** _____. The relatively stable electron

structures developed by loss of electrons in certain elements of groups 1B, 2B, 3A, and 4A

are called **(9)** _____.

For each statement below, write *true* or *false*.

_____ **10.** A positively charged ion is called an anion.

_____ **11.** Elements in group 1A lose their one valence electron, forming an ion with a 1+ charge.

_____ **12.** Elements tend to react so that they acquire the electron structure of a halogen.

_____ **13.** A sodium atom tends to lose one electron when it reacts.

_____ **14.** The electron structure of a zinc ion (Zn^{2+}) is an example of a pseudo-noble gas formation.

_____ **15.** A Cl^- ion is an example of a cation.

_____ **16.** The ending *-ide* is used to designate an anion.

_____ **17.** Nonmetals form a stable outer electron configuration by losing electrons and becoming anions.

Section 8.2 The Formation and Nature of Ionic Bonds

In your textbook, read about forming ionic bonds and the characteristics of ionic compounds.

Circle the letter of the choice that best completes the statement or answers the question.

1. An ionic bond is
   a. attraction of an atom for its electrons.
   b. attraction of atoms for electrons they share.
   c. a force that holds together atoms that are oppositely charged.
   d. the movement of electrons from one atom to another.

2. The formula unit of an ionic compound shows the
   a. total number of each kind of ion in a sample.
   b. simplest ratio of the ions.
   c. numbers of atoms within each molecule.
   d. number of nearest neighboring ions surrounding each kind of ion.

3. The overall charge of a formula unit for an ionic compound
   a. is always zero.
   c. is always positive.
   b. is always negative.
   d. may have any value.

4. How many chloride (Cl^-) ions are present in a formula unit of magnesium chloride, given that the charge on a Mg ion is 2+?
   a. one-half   b. one   c. two   d. four

5. Ionic bonds generally occur between
   a. metals.
   c. a metal and a nonmetal.
   b. nonmetals.
   d. noble gases.

6. Salts are examples of
   a. nonionic compounds.   b. metals.   c. nonmetals.   d. ionic compounds.

7. A three-dimensional arrangement of particles in an ionic solid is called a(n)
   a. crystal lattice.   b. sea of electrons.   c. formula unit.   d. electrolyte.

8. In a crystal lattice of an ionic compound,
   a. ions of a given charge are clustered together, far from ions of the opposite charge.
   b. ions are surrounded by ions of the opposite charge.
   c. a sea of electrons surrounds the ions.
   d. neutral molecules are present.

CHAPTER **8**   STUDY GUIDE FOR CONTENT MASTERY

Section 8.2 *continued*

9. What is the relationship between lattice energy and the strength of the attractive force holding ions in place?

  a. The more positive the lattice energy is, the greater the force.

  b. The more negative the lattice energy is, the greater the force.

  c. The closer the lattice energy is to zero, the greater the force.

  d. There is no relationship between the two quantities.

10. The formation of a stable ionic compound from ions

  a. is always exothermic.   **c.** is always endothermic.

  b. may be either exothermic or endothermic. **d.** neither absorbs nor releases energy.

11. In electron transfer involving a metallic atom and a nonmetallic atom during ion formation, which of the following is correct?

  a. The metallic atom gains electrons from the nonmetallic atom.

  b. The nonmetallic atom gains electrons from the metallic atom.

  c. Both atoms gain electrons.

  d. Neither atom gains electrons.

Underline the word that correctly describes each property in ionic compounds.

12. Melting point	Low	High
13. Boiling point	Low	High
14. Hardness	Hard	Soft
15. Brittleness	Flexible	Brittle
16. Electrical conductivity in the solid state	Good	Poor
17. Electrical conductivity in the liquid state	Good	Poor
18. Electrical conductivity when dissolved in water	Good	Poor

For each statement below, write *true* or *false*.

_____ **19.** The crystal lattice of ionic compounds affects their melting and boiling points.

_____ **20.** The lattice energy is the energy required to separate the ions of an ionic compound.

_____ **21.** The energy of an ionic compound is higher than that of the separate elements that formed it.

_____ **22.** Large ions tend to produce a more negative value for lattice energy than smaller ions do.

_____ **23.** Ions that have larger charges tend to produce a more negative lattice energy than ions with smaller charges do.

CHAPTER (8) STUDY GUIDE FOR CONTENT MASTERY

Section 8.3 Names and Formulas for Ionic Compounds

In your textbook, read about communicating what is in a compound and naming ions and ionic compounds.

Use each of the terms below just once to complete the passage.

anion	-ate	cation	electrons	zero
lower right	monatomic	one	oxidation number	-ite
oxyanion	polyatomic	subscript		

A one-atom ion is called a(n) **(1)**_____ ion. The charge of such an ion is

equal to the atom's **(2)**_____, which is the number of **(3)**_____

transferred to or from the atom to form the ion. In ionic compounds, the sum of the charges of

all the ions equals **(4)**_____. Ions made up of more than one atom are called

(5)_____ ions. If such an ion is negatively charged and includes one or more

oxygen atoms, it is called a(n) **(6)**_____. If two such ions can be formed that

contain different numbers of oxygen atoms, the name for the ion with more oxygen atoms ends

with the suffix **(7)**_____. The name for the ion with fewer oxygen atoms

ends with **(8)**_____.

In the chemical formula for any ionic compound, the chemical symbol for the

(9)_____ is written first, followed by the chemical symbol for the

(10)_____. A(n) **(11)**_____ is a small number used to

represent the number of ions of a given element in a chemical formula. Such numbers are

written to the **(12)**_____ of the symbol for the element. If no number

appears, the assumption is that the number equals **(13)**_____.

For each formula in Column A, write the letter of the matching name in Column B.

Column A	Column B
_____ **14.** ClO_2^-	**a.** chlorate
_____ **15.** ClO_4^-	**b.** hypochlorite
_____ **16.** ClO^-	**c.** chloride
_____ **17.** Cl^-	**d.** perchlorate
_____ **18.** ClO_3^-	**e.** chlorite

CHAPTER 8

STUDY GUIDE FOR CONTENT MASTERY

Section 8.3 *continued*

For each of the following chemical formulas, write the correct name of the ionic compound represented. You may refer to the periodic table on pages 156–157 and Table 8.7 for help.

19. NaI _____

20. $CaCl_2$ _____

21. K_2S _____

22. MgO _____

23. $LiHSO_4$ _____

24. NH_4Br _____

25. Ca_3N_2 _____

26. Cs_3P _____

27. $KBrO_3$ _____

28. $Mg(ClO)_2$ _____

29. Li_2O_2 _____

30. $Be_3(PO_4)_2$ _____

31. $(NH_4)_2CO_3$ _____

32. $NaBrO_3$ _____

33. Fe_2O_3 _____

34. $Fe(IO_3)_2$ _____

For each of the following ionic compounds, write the correct formula for the compound. You may refer to the periodic table on pages 156–157 and Table 8.7 for help.

35. beryllium nitride _____

36. nickel(II) chloride _____

37. potassium chlorite _____

38. copper(I) oxide _____

39. magnesium sulfite _____

40. ammonium sulfide _____

41. calcium iodate _____

42. iron(III) perchlorate _____

43. sodium nitride _____

CHAPTER 8 · STUDY GUIDE FOR CONTENT MASTERY

Section 8.4 Metallic Bonds and Properties of Metals

In your textbook, read about metallic bonds.

Use the diagram of metallic bonding to answer the following questions.

1. What is the name of the model of metallic bonding that is illustrated?

_____

2. Why are the electrons in a metallic solid described as delocalized?

_____

3. Which electrons from the metal make up the delocalized electrons?

_____

4. Are the metal atoms that are shown cations or anions? How can you tell?

_____

5. How do the metallic ions differ from the ions that exist in ionic solids?

_____

_____

6. Explain what holds the metal atoms together in the solid.

_____

_____

In your textbook, read about the properties of metals.

For each property, write *yes* if the property is characteristic of most metals, or *no* if it is not. If the property is a characteristic of metals, explain how metallic bonding accounts for the property.

7. Malleable _____

_____

8. Brittle _____

9. Lustrous _____

10. High melting point _____

11. Low boiling point _____

12. Ductile _____

_____

13. Poor conduction of heat _____

14. Good conduction of electricity _____

CHAPTER 9 STUDY GUIDE FOR CONTENT MASTERY

Covalent Bonding

Section 9.1 The Covalent Bond

In your textbook, read about the nature of covalent bonds.

Use each of the terms below just once to complete the passage.

covalent bond	molecule	sigma bond	exothermic	pi bond

When sharing of electrons occurs, the attachment between atoms that results is called

a(n) **(1)** _____. When such an attachment is formed, bond dissociation

energy is released, and the process is **(2)** _____. When two or more

atoms bond by means of electron sharing, the resulting particle is called a(n)

(3) _____. If the electrons shared are centered between the two atoms, the

attachment is called a(n) **(4)** _____. If the sharing involves the overlap of

parallel orbitals, the attachment is called a(n) **(5)** _____.

In your textbook, read about single and multiple bonds and bond strength.

Circle the letter of the choice that best completes the statement or answers the question.

6. In what form do elements such as hydrogen, nitrogen, and oxygen normally occur?

   a. as single atoms

   b. as molecules containing two atoms

   c. as molecules containing three atoms

   d. as molecules containing four atoms

7. How many electrons are shared in a double covalent bond?

   a. none **b.** one **c.** two **d.** four

8. Bond length is the distance between

   a. two molecules of the same substance.

   b. the electrons in two attached atoms.

   c. the nuclei of two attached atoms.

   d. the orbitals of two attached atoms.

9. Which of the following relationships relating to bond length is generally correct?

   a. the shorter the bond, the stronger the bond

   b. the shorter the bond, the weaker the bond

   c. the shorter the bond, the fewer the electrons in it

   d. the shorter the bond, the lower the bond dissociation energy

Section 9.2 Naming Molecules

In your textbook, read about how binary compounds and acids are named from their formulas.

For each statement below, write *true* or *false*.

_____ **1.** Binary molecular compounds are generally composed of a metal and a nonmetal.

_____ **2.** The second element in the formula of a binary compound is named using the suffix *-ite*.

_____ **3.** The prefix *tetra-* indicates three atoms.

_____ **4.** The prefix *hexa-* indicates six atoms.

_____ **5.** In naming the first element in a formula, the prefix *mono-* is not used.

_____ **6.** For binary acids, the hydrogen part of the compound is named using the prefix *hydro-*.

_____ **7.** An oxyacid contains only two elements.

_____ **8.** If the name of the anion of an oxyacid ends in *-ate*, the acid name contains the suffix *-ous*.

In your textbook, read about naming molecular compounds and oxyacids.

For each item in Column A, write the letter of the matching item in Column B.

Column A	Column B
_____ **9.** CO	**a.** hydrobromic acid
_____ **10.** CO_2	**b.** dinitrogen tetroxide
_____ **11.** H_2CO_3	**c.** carbon monoxide
_____ **12.** NH_3	**d.** nitrous acid
_____ **13.** N_2O_4	**e.** ammonia
_____ **14.** HNO_2	**f.** nitric acid
_____ **15.** HNO_3	**g.** carbonic acid
_____ **16.** HBr	**h.** bromic acid
_____ **17.** $HBrO_3$	**i.** carbon dioxide

Section 9.3 Molecular Structures

In your textbook, read about Lewis structures.

For each statement below, write *true* **or** *false*.

_____ **1.** A structural formula shows the arrangement of the atoms in a molecule.

_____ **2.** The central atom in a molecule is the one with the highest electron affinity.

_____ **3.** In molecules, hydrogen is always a terminal atom.

_____ **4.** The number of bonding pairs in a molecule is equal to the number of electrons.

_____ **5.** To find the total number of electrons available for bonding in a positive ion, you should add the ion charge to the total number of valence electrons of the atoms present.

_____ **6.** The electrons in a coordinate covalent bond are donated by both the bonded atoms.

_____ **7.** Resonance occurs when more than one valid Lewis structure can be written for a molecule.

_____ **8.** Nitrate is an example of an ion that forms resonance structures.

_____ **9.** The carbon dioxide molecule contains two double bonds.

_____ **10.** All electrons in an atom are available for bonding.

_____ **11.** In the sulfate ion (SO_4^{2-}), 32 electrons are available for bonding.

_____ **12.** When carbon and oxygen bond, the molecule contains ten pairs of bonding electrons.

In your textbook, read about resonance structures and exceptions to the octet rule.

For each item in Column A, write the letter of the matching item in Column B.

Column A	Column B
_____ **13.** Odd number of valence electrons	**a.** O_3
_____ **14.** Fewer than 8 electrons around an atom	**b.** BF_3
_____ **15.** More than 8 electrons around central atom	**c.** NO
_____ **16.** More than one valid Lewis structure	**d.** SF_6

Section 9.4 Molecular Shape

In your textbook, read about the VSEPR model.

Circle the letter of the choice that best completes the statement.

1. The VSEPR model is used mainly to

    a. determine molecular shape.
    c. determine ionic charge.

    b. write resonance structures.
    d. measure intermolecular distances.

2. The bond angle is the angle between

    a. the sigma and pi bonds in a double bond.
    c. two terminal atoms and the central atom.

    b. the nucleus and the bonding electrons.
    d. the orbitals of a bonding atom.

3. The VSEPR model is based on the idea that

    a. there is always an octet of electrons around an atom in a molecule.

    b. electrons are attracted to the nucleus.

    c. molecules repel one another.

    d. shared and unshared electron pairs repel each other as much as possible.

4. The shape of a molecule whose central atom has four pairs of bonding electrons is

    a. tetrahedral.
    b. trigonal planar.
    c. trigonal pyramidal.
    d. linear.

5. The shape of a molecule that has two covalent single bonds and no lone pairs on the central atom is

    a. tetrahedral.
    b. trigonal planar.
    c. trigonal pyramidal.
    d. linear.

6. The shape of a molecule that has three single covalent bonds and one lone pair on the central atom is

    a. tetrahedral.
    b. trigonal planar.
    c. trigonal pyramidal.
    d. linear.

In your textbook, read about hybridization.

Use each of the terms below just once to complete the passage.

carbon	hybridization	sp^3	identical	methane

    The formation of new orbitals from a combination or rearrangement of valence electrons

is called **(7)**_____. The orbitals that are produced in this way are

(8)_____ to one another. An example of an element that commonly

undergoes such formation is **(9)**_____. When this atom combines its three

p orbitals and its one s orbital, the orbitals that result are called **(10)**_____

orbitals. An example of a molecule that has this type of orbital is **(11)**_____.

Section 9.5 **Electronegativity and Polarity**

In your textbook, read about electronegativity.

Use the table of electronegativities below to answer the following questions.

Electronegativities of Some Elements

1 H 2.20	

| 3
Li
0.98 | 4
Be
1.57 |
| 11
Na
0.93 | 12
Mg
1.31 |

Metal
Metalloid
Nonmetal

| 5
B
2.04 | 6
C
2.55 | 7
N
3.04 | 8
O
3.44 | 9
F
3.98 |
| 13
Al
1.61 | 14
Si
1.90 | 15
P
2.19 | 16
S
2.58 | 17
Cl
3.16 |

19 K 0.82	20 Ca 1.00	21 Sc 1.36	22 Ti 1.54	23 V 1.63	24 Cr 1.66	25 Mn 1.55	26 Fe 1.83	27 Co 1.88	28 Ni 1.91	29 Cu 1.90	30 Zn 1.65	31 Ga 1.81	32 Ge 2.01	33 As 2.18	34 Se 2.55	35 Br 2.96
37 Rb 0.82	38 Sr 0.95	39 Y 1.22	40 Zr 1.33	41 Nb 1.6	42 Mo 2.16	43 Tc 2.10	44 Ru 2.2	45 Rh 2.28	46 Pd 2.20	47 Ag 1.93	48 Cd 1.69	49 In 1.78	50 Sn 1.96	51 Sb 2.05	52 Te 2.1	53 I 2.66
55 Cs 0.79	56 Ba 0.89	57 La 1.10	72 Hf 1.3	73 Ta 1.5	74 W 1.7	75 Re 1.9	76 Os 2.2	77 Ir 2.2	78 Pt 2.2	79 Au 2.4	80 Hg 1.9	81 Tl 1.8	82 Pb 1.8	83 Bi 1.9	84 Po 2.0	85 At 2.2
87 Fr 0.7	88 Ra 0.9	89 Ac 1.1														

1. What is the meaning of the term *electronegativity*?

_____

2. Which element has the highest electronegativity? What is the numerical value? What are the name and group number of the chemical family that has the highest overall electronegativities?

_____

3. Which element has the lowest electronegativity? What is the numerical value? What are the name and group number of the chemical family that has the lowest overall electronegativities?

_____

4. What general trend in electronegativity do you note going down a group? Across a period?

_____

5. How are the electronegativity values used to determine the type of bond that exists between two atoms?

_____

In your textbook, read about the properties of covalent compounds.

For each statement below, write *true* or *false*.

_____ **6.** Ionic compounds are usually soluble in polar substances.

_____ **7.** In a covalent molecular compound, the attraction between molecules tends to be strong.

CHAPTER **9** STUDY GUIDE FOR CONTENT MASTERY

Section 9.5 *continued*

In your textbook, read about bond polarity.

Using the table of electronegativities on the preceding page, circle the letter of the choice that best completes the statement or answers the question.

8. Unequal sharing of electrons between two bonded atoms always indicates

 a. a nonpolar covalent bond. **c.** a polar covalent bond.

 b. an ionic bond. **d.** a polar molecule.

9. When electronegativities of two bonded atoms differ greatly, the bond is

 a. polar covalent. **b.** coordinate covalent. **c.** polar covalent. **d.** ionic.

10. What is the electronegativity difference that usually is the dividing line between covalent and ionic bonds?

 a. 1.0 **b.** 1.7 **c.** 2.7 **d.** 4.0

11. The symbol δ^+ is placed next to which of the following?

 a. the less electronegative atom in a polar covalent bond **c.** a positive ion

 b. the more electronegative atom in a polar covalent bond **d.** the nucleus

12. A nonpolar covalent bond is one in which

 a. electrons are transferred. **c.** electrons are shared equally.

 b. electrons are shared unequally. **d.** both electrons are provided by the same atom.

13. Molecules containing only polar covalent bonds

 a. are always polar. **c.** are always ionic.

 b. may or may not be polar. **d.** are always nonpolar.

14. What factor other than electronegativity determines whether a molecule as a whole is polar or not?

 a. temperature **b.** its geometry **c.** its physical state **d.** its mass

15. Which of the following correctly describes the compound water, H_2O?

 a. ionic **c.** polar overall, with nonpolar covalent bonds

 b. nonpolar overall, with polar covalent bonds **d.** polar overall, with polar covalent bonds

16. Which of the following correctly describes the compound carbon tetrachloride, CCl_4?

 a. ionic **c.** polar overall, with nonpolar covalent bonds

 b. nonpolar overall, with polar covalent bonds **d.** polar overall, with polar covalent bonds

17. A molecule of ammonia, NH_3, is

 a. nonpolar because it is linear.

 b. polar because it is linear.

 c. nonpolar because there is no electronegativity difference.

 d. polar because there is an electronegativity difference and the molecule is trigonal pyramidal.

CHAPTER 10 STUDY GUIDE FOR CONTENT MASTERY

Chemical Reactions

Section 10.1 Reactions and Equations

In your textbook, read about evidence of chemical reactions.

For each statement, write *yes* if evidence of a chemical reaction is present. Write *no* if there is no evidence of a chemical reaction.

_____ **1.** A tomato smells rotten.

_____ **2.** A drinking glass breaks into smaller pieces.

_____ **3.** A piece of ice melts.

_____ **4.** Drain cleaner is mixed with water and the solution becomes warm.

_____ **5.** Candle wax burns.

_____ **6.** Molten candle wax solidifies.

_____ **7.** Green leaves turn yellow and red as the seasons change.

_____ **8.** Baking powder produces a gas that makes a cake rise.

In your textbook, read about how to represent chemical reactions and how to balance chemical equations.

Use the terms below to complete the passage. Each term may be used once, more than once, or not at all.

arrow	plus sign	(s)	(l)
reactant	product	(g)	(aq)

The fuel for the space shuttle is hydrogen, which burns in oxygen to produce water vapor and energy. In this chemical reaction, hydrogen is a(n) **(9)** _____, oxygen is a(n) **(10)** _____, and water vapor is a(n) **(11)** _____. In a chemical equation for this reaction, a(n) **(12)** _____ is used to separate hydrogen and oxygen from water vapor and energy. A(n) **(13)** _____ is used to separate the symbols for hydrogen and oxygen. A(n) **(14)** _____ symbol is used to tell the state of hydrogen in the reaction, a(n) **(15)** _____ symbol is used for the state of oxygen, and a(n) **(16)** _____ symbol is used for the state of water vapor.

Section 10.1 *continued*

For each of the following chemical reactions, write a word equation, a skeleton equation, and a balanced chemical equation. Be sure to show the state of each reactant and product. If you need more help writing formulas or determining the state of a substance, refer to Chapters 8 and 9 and the periodic table on pages 156–157.

17. Solid mercury(II) oxide breaks down when heated, forming the elements mercury and oxygen.

_____

_____

18. Sodium metal reacts with water vapor in air to form solid sodium hydroxide and hydrogen.

_____

_____

19. In the first step of refining zinc metal from its zinc sulfide ore, the ore is heated in the presence of oxygen. The products are solid zinc oxide and sulfur dioxide gas.

_____

_____

20. The next step of refining zinc involves heating the zinc oxide in the presence of carbon. This reaction produces zinc vapor and carbon monoxide gas.

_____

_____

21. Certain pollutants in the air react with water vapor to form acids. For example, sulfur trioxide reacts with water vapor to form sulfuric acid.

_____

_____

22. Solid calcium carbonate is commonly used in antacids because it reacts with the hydrochloric acid found in the stomach. The products of this reaction are aqueous calcium chloride, carbon dioxide, and water.

_____

_____

_____

CHAPTER 10

Section 10.2 Classifying Chemical Reactions

In your textbook, read about synthesis, combustion, decomposition, and replacement reactions.

Assume that Q, T, X, and Z are symbols for elements. Match each equation in Column A with the reaction type it represents in Column B.

Column A	Column B
_____ **1.** Q + XZ → X + QZ	**a.** decomposition
_____ **2.** Q + Z → QZ	**b.** double-replacement
_____ **3.** QT → Q + T	**c.** single-replacement
_____ **4.** QT + XZ → QZ + XT	**d.** synthesis

Answer the following questions.

5. Does the following equation represent a combustion reaction, a synthesis reaction, or both? Explain your answer.

$$2C(s) + O_2(g) \rightarrow 2CO_2(g) + energy$$

_____

_____

_____

6. Why is it sometimes incorrect to state that a compound is broken down into its component elements in a decomposition reaction?

_____

_____

_____

7. When soap is added to hard water, solid soap scum forms. When water is added to baking powder, carbon dioxide bubbles form. When lemon juice is added to household ammonia solution, water is one of the products. Tell how you know a double-replacement reaction has occurred in each case.

_____

_____

_____

8. Explain how you can use an activity series to determine whether a single-replacement reaction will occur.

_____

_____

_____

Section 10.2 *continued*

In your textbook, read about the activity series for metal and halogens.

Examine each of the following pairs of potential reactants. Use Figure 10-10 in your textbook to help you decide whether or not a reaction would occur. If a reaction occurs, write the balanced equation. If no reaction occurs, write *NR*.

9. calcium and water _____

10. magnesium and water _____

11. rubidium and lithium chloride _____

12. potassium and aluminum oxide _____

13. silver and calcium nitrate _____

14. fluorine and potassium iodide _____

15. magnesium bromide and chlorine _____

16. copper and iron(III) sulfate _____

Match each example of a chemical reaction in Column A to the type(s) listed in Column B. List all types from Column B that apply.

Column A	Column B
_____ **17.** Aluminum lawn furniture becomes coated with a layer of aluminum oxide when it sits out in the air.	**a.** combustion
_____ **18.** Chlorine gas is bubbled through a calcium bromide solution. The solution turns brown, the color of bromine.	**b.** decomposition
_____ **19.** Lime is added to acid water in a lake. Water and a salt form.	**c.** double-replacement
_____ **20.** Propane is a common household fuel. When burned, water and carbon dioxide are produced.	**d.** single-replacement
_____ **21.** Steel wool burns, forming an iron oxide.	**e.** synthesis
_____ **22.** When an electric current is passed through molten potassium bromide, potassium and bromine form.	
_____ **23.** When solutions of sodium iodide and lead nitrate are combined, a yellow solid forms.	

CHAPTER 10

STUDY GUIDE FOR CONTENT MASTERY

Section 10.3 Reactions in Aqueous Solutions

In your textbook, read about aqueous solutions, reactions that form precipitates, reactions that form water, and reactions that form gases.

Circle the letter of the choice that best completes the statement or answers the question.

1. A spoonful of sodium chloride is dissolved in a liter of water. What is sodium chloride in this solution?

   a. molecule   **b.** precipitate   **c.** solute   **d.** solvent

2. In an aqueous solution, water is the

   a. homogeneous part.   **b.** precipitate.   **c.** solute.   **d.** solvent.

3. Compounds that produce hydrogen ions in aqueous solutions are

   a. acids.   **b.** aqueous.   **c.** bases.   **d.** ionic compounds.

4. What type of reaction occurs between ions present in aqueous solution?

   a. decomposition   **b.** double-replacement   **c.** single-replacement   **d.** synthesis

5. What type of ions are present in solution but are not actually involved in a chemical reaction?

   a. complete   **b.** net   **c.** precipitate   **d.** spectator

6. If hydrochloric acid and potassium hydroxide react, what is the product of the net ionic equation for the reaction?

   a. hydrochloric acid   **b.** hydrogen ions   **c.** potassium chloride   **d.** water

7. Which of the following gases is not commonly produced in a double-replacement reaction?

   a. carbon dioxide   **b.** hydrogen cyanide   **c.** hydrogen sulfide   **d.** sulfur dioxide

8. $H^+(aq) + Br^-(aq) + K^+(aq) + OH^-(aq) \rightarrow H_2O(l) + Br^-(aq) + K^+(aq)$ is an example of what type of chemical equation?

   a. complete ionic   **b.** net ionic   **c.** precipitation   **d.** spectator

Section 10.3 *continued*

Predict the products for each reaction in Column A. Write the formulas for these products on the product side of each equation. In the space at the left, write the letter of the choice from Column B that indicates what type of product is produced during the reaction shown in Column A. Write as many choices as apply. (Hints: Compounds of Group 1 metals are never precipitates; H_2S and CO_2 are gases.)

Column A	Column B
_____ **9.** $HBr(aq) + KOH(aq) \rightarrow$ _____	**a.** gas
_____ **10.** $HNO_3(aq) + Na_2CO_3(aq) \rightarrow$ _____	**b.** precipitate
_____ **11.** $NaI(aq) + Pb(C_2H_3O_2)_2(aq) \rightarrow$ _____	**c.** water
_____ **12.** $CsOH(aq) + H_2SO_4(aq) \rightarrow$ _____	
_____ **13.** $K_2S(aq) + HCl(aq) \rightarrow$ _____	

For each of the following reactions, write chemical, complete ionic, and net ionic equations.

14. Phosphoric acid (H_3PO_4) and lithium hydroxide react to form a salt and water.

_____

_____

15. When solutions of magnesium sulfate and calcium chloride are mixed, calcium sulfate precipitates.

_____

_____

_____

16. Bubbles are released when nitric acid (HNO_3) is added to a potassium carbonate solution.

_____

_____

_____

17. Bubbles are released when hydrobromic acid (HBr) is added to a solution of ammonium sulfide. Aqueous ammonium bromide also forms.

_____

_____

CHAPTER 11 STUDY GUIDE FOR CONTENT MASTERY

The Mole

Section 11.1 Measuring Matter

In your textbook, read about counting particles.

In Column B, rank the quantities from Column A from smallest to largest.

Column A	Column B
0.5 mol	**1.** _____
200	**2.** _____
5	**3.** _____
6 000 000 000	**4.** _____
6.02×10^{23}	**5.** _____
dozen	**6.** _____
four moles	**7.** _____
gross	**8.** _____
pair	**9.** _____
ream	**10.** _____

In your textbook, read about converting moles to particles and particles to moles.

In the boxes provided, write the conversion factor that correctly completes each problem.

11. 1.20 mol Cu $\times$ [   ] $= 7.22 \times 10^{23}$ Cu atoms

12. 9.25×10^{22} molecules CH_4 $\times$ [   ] $= 1.54 \times 10^{-1}$ mol CH_4

13. 1.54×10^{26} atoms Xe $\times$ [   ] $= 2.56 \times 10^2$ mol Xe

14. 3.01 mol F_2 $\times$ [   ] $= 1.81 \times 10^{24}$ molecules F_2

CHAPTER 11   **STUDY GUIDE FOR CONTENT MASTERY**

Section 11.2 Mass and the Mole

In your textbook, read about the mass of a mole.

For each statement below, write *true* or *false*.

_____ **1.** The isotope hydrogen-1 is the standard used for the relative scale of atomic masses.

_____ **2.** The mass of an atom of helium-4 is 4 amu.

_____ **3.** The mass of a mole of hydrogen atoms is 1.00×10^{23} amu.

_____ **4.** The mass in grams of one mole of any pure substance is called its molar mass.

_____ **5.** The atomic masses recorded on the periodic table are weighted averages of the masses of all the naturally occurring isotopes of each element.

_____ **6.** The molar mass of any element is numerically equal to its atomic mass in grams.

_____ **7.** The molar mass unit is mol/g.

_____ **8.** If the measured mass of an element is numerically equal to its molar mass, then you have indirectly counted 6.02×10^{23} atoms of the element in the measurement.

In your textbook, read about using molar mass.

For each problem listed in Column A, select from Column B the letter of the conversion factor that is needed to solve the problem. You may need to use more than one conversion factor to solve the problem.

Column A	Column B
_____ **9.** Find the number of moles in 23.0 g of zinc.	**a.** $\dfrac{65.4 \text{ g Zn}}{1 \text{ mol Zn}}$
_____ **10.** Find the mass of 5.0×10^{20} zinc atoms.	
_____ **11.** Find the mass of 2.00 moles of zinc.	**b.** $\dfrac{1 \text{ mol Zn}}{65.4 \text{ g Zn}}$
_____ **12.** Find the number of atoms in 7.40 g of zinc.	
_____ **13.** Find the number of moles that contain 4.25×10^{27} zinc atoms.	**c.** $\dfrac{6.02 \times 10^{23} \text{ atoms Zn}}{1 \text{ mol Zn}}$
_____ **14.** Find the number of atoms in 3.25 moles of zinc.	**d.** $\dfrac{1 \text{ mol Zn}}{6.02 \times 10^{23} \text{ atoms Zn}}$

Section 11.3 Moles of Compounds

In your textbook, read about chemical formulas and the mole, the molar mass of compounds, and conversions among mass, moles, and number of particles.

Study the table and the diagram of a methane molecule and a trichloromethane molecule. Then answer the following questions.

Element	Molar Mass (g/mol)
Hydrogen	1.01
Carbon	12.01
Chlorine	35.45

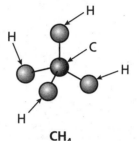

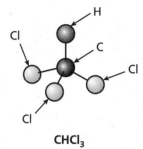

CH_4 $CHCl_3$

1. What elements and how many atoms of each does a molecule of methane contain?

_____

2. What elements and how many atoms of each does a molecule of trichloromethane contain?

_____

3. How many moles of each element are in a mole of methane?

_____

4. How many moles of each element are in a mole of trichloromethane?

_____

5. Which of the following values represents the number of carbon atoms in one mole of methane? 6.02×10^{23}; 12.0×10^{23}; 18.1×10^{23}; 24.1×10^{23}

_____

6. Which of the following values represents the number of chlorine atoms in one mole of trichloromethane? 6.02×10^{23}; 1.20×10^{24}; 1.81×10^{24}; 2.41×10^{23}

_____

7. Which of the following values represents the molar mass of methane? 13.02 g/mol; 16.05 g/mol; 52.08 g/mol; 119.37 g/mol

_____

8. Chloromethane (CH_3Cl) has a molar mass of 50.49 g/mol. Which of the following values represents the number of molecules of CH_3Cl in 101 grams of the substance? 3.01×10^{23}; 6.02×10^{23}; 1.20×10^{24}; 6.08×10^{26}

_____

CHAPTER 11 STUDY GUIDE FOR CONTENT MASTERY

Section 11.4 Empirical and Molecular Formulas

In your textbook, read about percent composition.

Answer the following questions.

1. What is the percent composition of a compound?

_____

_____

2. Describe how to find the percent composition of a compound if you know the mass of a sample of a compound and the mass of each element in the sample.

_____

_____

In your textbook, read about empirical and molecular formulas.

Circle the letter of the choice that best answers the question.

3. Which information about a compound can you use to begin to determine the empirical and molecular formulas of the compound?

   a. mass of the compound   **c.** percent composition of the compound

   b. number of elements in the compound   **d.** volume of the compound

4. You have determined that a compound is composed of 0.300 moles of carbon and 0.600 moles of oxygen. What must you do to determine the mole ratio of the elements in the empirical formula of the compound?

   a. Multiply each mole value by 0.300 mol.   **c.** Divide each mole value by 0.300 mol.

   b. Multiply each mole value by 0.600 mol.   **d.** Divide each mole value by 0.600 mol.

5. The mole ratio of carbon to hydrogen to oxygen in a compound is 1 mol C : 2 mol H : 1 mol O. What is the empirical formula of the compound?

   a. CHO   **b.** CH_2O **c.** C_2HO_2 **d.** $C_2H_2O_2$

6. You calculate the mole ratio of oxygen to aluminum in a compound to be 1.5 mol O : 1 mol Al. What should you do to determine the mole ratio in the empirical formula of the compound?

   a. Multiply each mole value by 1.5.   **c.** Divide each mole value by 1.5.

   b. Multiply each mole value by 2.   **d.** Divide each mole value by 2.

7. What is the relationship between the molecular formula and the empirical formula of a compound?

   a. (molecular formula)(empirical formula) = n

   b. molecular formula = $\dfrac{\text{empirical formula}}{n}$

   c. molecular formula = (empirical formula)n

   d. molecular formula = $\dfrac{n}{\text{empirical formula}}$

8. You know that the empirical formula of a compound has a molar mass of 30.0 g/mol. The experimental molar mass of this compound is 60.0 g/mol. What must you do to determine the value of n in the relationship between the molecular formula and the empirical formula?

   a. Add 30.0 g/mol and 60.0 g/mol.   **c.** Divide 60.0 g/mol by 30.0 g/mol.

   b. Divide 30.0 g/mol by 60.0 g/mol.   **d.** Multiply 30.0 g/mol by 60.0 g/mol.

9. You know that the experimental molar mass of a compound is three times the molar mass of its empirical formula. If the compound's empirical formula is NO_2, what is its molecular formula?

   a. NO_2 **b.** NO_6 **c.** N_3O_2 **d.** N_3O_6

Solve the following problem. Show your work in the space provided.

10. A sample of a compound contains 7.89 g potassium, 2.42 g carbon, and 9.69 g oxygen. Determine the empirical and molecular formulas of this compound, which has a molar mass of 198.22 g/mol.

Section 11.5 The Formula for a Hydrate

In your textbook, read about naming and analyzing hydrates.

Use each of the terms below just once to complete the passage.

anhydrous	crystal structure	desiccants	formula unit
hydrate	hydration	water molecules	water of hydration

A(n) **(1)**_____ is a compound that has a specific number of water

molecules bound to its atoms. Molecules of water that become part of a hydrate are called

waters of **(2)**_____. In the formula for a hydrate, the number of

(3)_____ associated with each **(4)**_____ of the

compound is written following a dot.

The substance remaining after a hydrate has been heated and its waters of hydration

released is called **(5)**_____. The ratio of the number of moles of

(6)_____ to one mole of the anhydrous compound indicates the

coefficient of H_2O that follows the dot in the formula of the hydrate. Because the anhydrous

form of the hydrate can absorb water into its **(7)**_____, hydrates are used

as **(8)**_____, which are drying agents.

Complete the table of hydrates.

Chemical Formula	Name
$CdSO_4$	Cadmium sulfate, anhydrous
$CdSO_4 \cdot H_2O$	**9.**
10.	Cadmium sulfate tetrahydrate

Solve the following problem. Show your work in the space provided.

11. A 2.00-g sample of a hydrate of iron(II) chloride produces 1.27 g of anhydrous iron(II) chloride ($FeCl_2$) after heating. Determine the empirical formula and the name of the hydrate.

CHAPTER 12 STUDY GUIDE FOR CONTENT MASTERY

Stoichiometry

Section 12.1 What is stoichiometry?

In your textbook, read about stoichiometry and the balanced equation.

For each statement below, write *true* or *false*.

_____ **1.** The study of the quantitative relationships between the amounts of reactants used and the amounts of products formed by a chemical reaction is called stoichiometry.

_____ **2.** Stoichiometry is based on the law of conservation of mass.

_____ **3.** In any chemical reaction, the mass of the products is less than the mass of the reactants.

_____ **4.** The coefficients in a chemical equation represent not only the number of individual particles but also the number of moles of particles.

_____ **5.** The mass of each reactant and product is related to its coefficient in the balanced chemical equation for the reaction by its molar mass.

Complete the table below, using information represented in the chemical equation for the combustion of methanol, an alcohol.

$$\text{methanol} + \text{oxygen} \rightarrow \text{carbon dioxide} + \text{water}$$

$$2CH_3OH(l) + 3O_2(g) \rightarrow 2CO_2(g) + 4H_2O(g)$$

Substance	Molar Mass (g/mol)	Number of Molecules	Number of Moles (mol)	Mass (g)
6. Methanol	32.05			
7. Oxygen gas	32.00			
8. Carbon dioxide	44.01			
9. Water	18.02			

10. What are the reactants? _____

11. What are the products? _____

12. What is the total mass of the reactants? _____

13. What is the total mass of the products? _____

14. How do the total masses of the reactants and products compare? _____

CHAPTER 12   **STUDY GUIDE FOR CONTENT MASTERY**

Section 12.1 *continued*

In your textbook, read about mole ratios.

Answer the questions about the following chemical reaction.

$$\text{sodium} + \text{iron(III) oxide} \rightarrow \text{sodium oxide} + \text{iron}$$

$$6Na(s) + Fe_2O_3(s) \rightarrow 3Na_2O(s) + 2Fe(s)$$

15. What is a mole ratio?

_____

_____

16. How is a mole ratio written?

_____

_____

_____

_____

17. Predict the number of mole ratios for this reaction. _____

18. What are the mole ratios for this reaction?

19. What is the mole ratio relating sodium to iron? _____

20. What is the mole ratio relating iron to sodium? _____

21. Which mole ratio has the largest value? _____

Section 12.2 Stoichiometric Calculations

In your textbook, read about mole-to-mole conversion.

Read the following passage and then solve the problems. In the equation that follows each problem, write in the space provided the mole ratio that can be used to solve the problem. Complete the equation by writing the correct value on the line provided.

The reaction of sodium peroxide and water produces sodium hydroxide and oxygen gas. The following balanced chemical equation represents the reaction.

$$2Na_2O_2(s) + 2H_2O(l) \rightarrow 4NaOH(s) + O_2(g)$$

1. How many moles of sodium hydroxide are produced when 1.00 mol sodium peroxide reacts with water?

1.00 mol Na_2O_2 × _____ = _____ mol NaOH

2. How many moles of oxygen gas are produced when 0.500 mol Na_2O_2 reacts with water?

0.500 mol Na_2O_2 × _____ = _____ mol O_2

3. How many moles of sodium peroxide are needed to produce 1.00 mol sodium hydroxide?

1.00 mol NaOH × _____ = _____ mol Na_2O_2

4. How many moles of water are required to produce 2.15 mol oxygen gas in this reaction?

2.15 mol O_2 × _____ = _____ mol H_2O

5. How many moles of water are needed for 0.100 mol of sodium peroxide to react completely in this reaction?

0.100 mol Na_2O_2 × _____ = _____ mol H_2O

6. How many moles of oxygen are produced if the reaction produces 0.600 mol sodium hydroxide?

0.600 mol NaOH × _____ = _____ mol O_2

Section 12.2 *continued*

In your textbook, read about mole-to-mass and mass-to-mass conversions.

Solving a mass-to-mass problem requires the four steps listed below. The equations in the boxes show how the four steps are used to solve an example problem. After you have studied the example, solve the problems below, using the four steps.

Example problem: How many grams of carbon dioxide are produced when 20.0 g acetylene (C_2H_2) is burned?

Solution
$2C_2H_2(g) + 5O_2(g)$ $\rightarrow 4CO_2(g) + 2H_2O(g)$
$20.0 \text{ g } C_2H_2 \times \dfrac{1 \text{ mol } C_2H_2}{26.04 \text{ g } C_2H_2}$ $= 0.768 \text{ mol } C_2H_2$
$0.768 \text{ mol } C_2H_2 \times \dfrac{4 \text{ mol } CO_2}{2 \text{ mol } C_2H_2}$ $= 1.54 \text{ mol } CO_2$
$1.54 \text{ mol } CO_2 \times \dfrac{44.01 \text{ g } CO_2}{1 \text{ mol } CO_2}$ $= 67.8 \text{ g } CO_2$

Step 1 Write a balanced chemical equation for the reaction.

Step 2 Determine the number of moles of the known substance, using mass-to-mole conversion.

Step 3 Determine the number of moles of the unknown substance, using mole-to-mole conversion.

Step 4 Determine the mass of the unknown substance, using mole-to-mass conversion.

7. In some mole-to-mass conversions, the number of moles of the known substance is given. In those conversions, which step of the above solution is not necessary? _____

8. In a blast furnace, iron and carbon monoxide are produced from the reaction of iron(III) oxide (Fe_2O_3) and carbon. How many grams of iron are formed when 150 g iron(III) oxide reacts with an excess of carbon?

9. Solid sulfur tetrafluoride (SF_4) and water react to form sulfur dioxide and an aqueous solution of hydrogen fluoride. How many grams of water are necessary for 20.0 g sulfur tetrafluoride to react completely?

Section 12.3 **Limiting Reactants**

In your textbook, read about why reactions stop and how to determine the limiting reactant.

Study the diagram showing a chemical reaction and the chemical equation that represents the reaction. Then complete the table. Show your calculations for questions 25–27 in the space below the table.

$$O_2 + 2NO \rightarrow 2NO_2$$

The molar masses of O_2, NO, and NO_2 are 32.00 g/mol, 30.01 g/mol, and 46.01 g/mol, respectively.

Amount of O_2	Amount of NO	Amount of NO_2	Limiting Reactant	Amount and Name of Excess Reactant
1 molecule	2 molecules	2 molecules	none	none
4 molecules	4 molecules	4 molecules	NO	2 molecules O_2
2 molecules	8 molecules	**1.**	**2.**	**3.**
1.00 mol	2.00 mol	**4.**	**5.**	**6.**
4.00 mol	4.00 mol	**7.**	**8.**	**9.**
5.00 mol	7.00 mol	**10.**	**11.**	**12.**
1.00 mol	4.00 mol	**13.**	**14.**	**15.**
0.500 mol	0.200 mol	**16.**	**17.**	**18.**
32.00 g	60.02 g	**19.**	**20.**	**21.**
16.00 g	80.00 g	**22.**	**23.**	**24.**
10.00 g	20.00 g	**25.**	**26.**	**27.**

Section 12.4 Percent Yield

In your textbook, read about the yields of products.

Study the diagram and the example problem.

$$\text{percent yield} = \frac{\text{actual yield}}{\text{theoretical yield}} \times 100\%$$

| mass of product from experimental measurement |

mass of product predicted from stoichiometric calculation using
a. mass of reactant
b. 4-step mass-to-mass conversion
  1. Write the balanced chemical equation.
  2. Calculate the number of moles of reactant, using molar mass.
  3. Calculate the number of moles of product, using the appropriate mole ratio.
  4. Calculate the mass of product, using the reciprocal of molar mass.

Example Problem: The following chemical equation represents the production of gallium oxide, a substance used in the manufacturing of some semiconductor devices.

$$4Ga(s) + 3O_2(g) \rightarrow 2Ga_2O_3(s)$$

In one experiment, the reaction yielded 7.42 g of the oxide from a 7.00-g sample of gallium. Determine the percent yield of this reaction. The molar masses of Ga and Ga_2O_3 are 69.72 g/mol and 187.44 g/mol, respectively.

Use the information in the diagram and example problem to evaluate each value or expression below. If the value or expression is correct, write *correct*. If it is incorrect, write the correct value or expression.

1. actual yield: unknown _____

2. mass of reactant: 7.00 g Ga _____

3. number of moles of reactant: $7.00 \text{ g Ga} \times \dfrac{69.72 \text{ g Ga}}{1 \text{ mol Ga}}$ _____

4. number of moles of product: $0.100 \text{ mol Ga} \times \dfrac{2 \text{ mol Ga}_2O_3}{1 \text{ mol Ga}}$ _____

5. theoretical yield: $0.0500 \text{ mol Ga}_2O_3 \times \dfrac{187.44 \text{ g Ga}_2O_3}{1 \text{ mol Ga}_2O_3}$ _____

6. percent yield: $\dfrac{9.37 \text{ g Ga}_2O_3}{7.42 \text{ g Ga}_2O_3} \times 100$ _____

CHAPTER 13 STUDY GUIDE FOR CONTENT MASTERY

States of Matter

Section 13.1 Gases

In your textbook, read about the kinetic-molecular theory.

Complete each statement.

1. The kinetic molecular theory describes the behavior of gases in terms of particles in

_____.

2. The kinetic-molecular theory makes the following assumptions.

  a. In a sample of a gas, the volume of the gas particles themselves is very

  _____ compared to the volume of the sample.

  b. Because gas particles are far apart, there are no significant attractive or repulsive

  _____ between gas particles.

  c. Gas particles are in constant and _____ motion.

  d. The collisions between gas particles are _____; that is, no

  _____ energy is lost.

3. The kinetic energy of a particle is represented by the equation _____.

4. _____ is a measure of the average kinetic energy of the particles in a
sample of matter.

In your textbook, read about explaining the behavior of gases.

For each statement below, write *true* or *false*.

_____ **5.** Gases are less dense than solids because there is a lot of space between the
particles of a gas.

_____ **6.** The random motion of gas particles causes a gas to expand until it fills its container.

_____ **7.** The density of a gas decreases as it is compressed.

_____ **8.** A gas can flow into a space occupied by another gas.

_____ **9.** The diffusion of a gas is caused by the random motion of the particles of the gas.

_____ **10.** Lighter gas particles diffuse less rapidly than do heavier gas particles.

_____ **11.** During effusion. a gas escapes through a tiny opening into a vacuum.

_____ **12.** Graham's law of effusion states that the rate of effusion for a gas is
directly related to the square root of its molar mass.

CHAPTER 13   **STUDY GUIDE FOR CONTENT MASTERY**

Section 13.1 *continued*

In your textbook, read about gas pressure.

Circle the letter of the choice that best completes the statement or answers the question.

13. Pressure is defined as force per unit

    a. area.   **b.** mass.   **c.** time.   **d.** volume.

14. What is an instrument designed to measure atmospheric pressure?

    a. barometer   **b.** manometer   **c.** sphygmomanometer **d.** thermometer

15. The height of the liquid in a barometer is affected by all of the following EXCEPT the

    a. altitude.   **c.** density of the liquid in the column.

    b. atmospheric pressure.   **d.** diameter of the column tube.

16. The pressure of the gas in a manometer is directly related to which of the following quantities?

    a. height of the mercury column in the closed-end arm

    b. height of the mercury column in the open-end arm

    c. a + b

    d. a − b

17. One atmosphere is equal to a pressure of

    a. 76 mm Hg.   **b.** 101.3 kPa.   **c.** 147 psi.   **d.** 706 torr.

18. The partial pressure of a gas depends on all of the following EXCEPT the

    a. concentration of the gas.   **c.** size of the container.

    b. identity of the gas.   **d.** temperature of the gas.

19. The pressure of a sample of air in a manometer is 102.3 kPa. What is the partial pressure of nitrogen (N_2) in the sample if the combined partial pressures of the other gases is 22.4 kPa?

    a. 62.4 kPa   **b.** 79.9 kPa   **c.** 102.3 kPa.   **d.** 124.7 kPa

Use the figure to answer the following questions.

20. What instrument is illustrated in the figure? _____

21. Who invented this instrument? _____

22. What are the two opposing forces that control the height of the mercury in the column?

_____

23. What does it mean when the level of mercury rises in the column?

_____

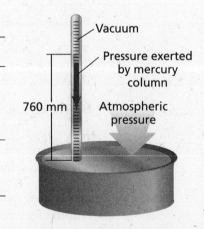

CHAPTER 13

STUDY GUIDE FOR CONTENT MASTERY

Section 13.2 Forces of Attraction

In your textbook, read about forces of attraction.

Answer the following questions.

1. Ionic, metallic, and covalent bonds are examples of what type of forces? _____

2. Dispersion forces, dipole–dipole forces, and hydrogen bonds are examples of what type
   of forces? _____

3. Describe dispersion forces.

   _____

   _____

4. Dispersion forces are greatest between what type of molecules?

   _____

5. Describe a permanent dipole.

   _____

   _____

6. Describe dipole–dipole forces.

   _____

   _____

7. Describe a hydrogen bond.

   _____

   _____

   _____

8. Identify each of the diagrams below as illustrating dipole–dipole forces, dispersion
   forces, or hydrogen bonds.

   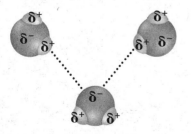   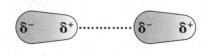   

   a. _____   **b.** _____   **c.** _____

9. Rank dipole-dipole forces, dispersion forces, and hydrogen bonds in order of increasing strength.

   _____

CHAPTER 13 STUDY GUIDE FOR CONTENT MASTERY

Section 13.3 Liquids and Solids

In your textbook, read about liquids and solids.

In the space at the left, write *true* if the statement is true; if the statement is false, change the italicized word or phrase to make it true.

_____ **1.** The constant *motion* of the particles in a liquid causes the liquid to take the shape of its container.

_____ **2.** At room temperature and one atmosphere of air pressure, the density of a liquid is much *greater* than that of its vapor.

_____ **3.** Liquids are not easily compressed because their particles are *loosely* packed.

_____ **4.** A liquid is less fluid than a gas because *intramolecular* attractions interfere with the ability of particles to flow past one another.

_____ **5.** Liquids that have stronger intermolecular forces have *higher* viscosities than do liquids with weaker intermolecular forces.

_____ **6.** The viscosity of a liquid *increases* with temperature because the increased average kinetic energy of the particles makes it easier for the particles to flow.

_____ **7.** Liquids that can form hydrogen bonds generally have a *high* surface tension.

_____ **8.** A liquid that rises in a narrow glass tube shows that the adhesive forces between the particles of the liquid and glass are *greater* than the cohesive forces between the particles of the liquid.

_____ **9.** Solids have a definite shape and volume because the motion of their particles is limited to *vibrations* around fixed locations.

_____ **10.** Most solids are *less* dense than liquids because the particles in a solid are more closely packed than those in a liquid.

_____ **11.** Rubber is *a crystalline* solid because its particles are not arranged in a regular, repeating pattern.

Section 13.4 Phase Changes

In your textbook, read about phase changes.

Complete the table by writing the initial and final phases for each phase change and making a check (✔) in the correct energy column.

Phase Change	Phase		Energy	
	initial	final	required	released
1. Condensation				
2. Deposition				
3. Freezing				
4. Melting				
5. Sublimation				
6. Vaporization				

For each item in Column A, write the letter of the matching item in Column B.

Column A	Column B
_____ **7.** Temperature at which a liquid is converted into a crystalline solid	**a.** boiling point
_____ **8.** Temperature at which the forces holding a crystalline lattice together are broken	**b.** freezing point
_____ **9.** Temperature at which the vapor pressure of a liquid equals the external or atmospheric pressure	**c.** melting point

Section 13.4 *continued*

In your textbook, read about phase diagrams.

Use the phase diagram for water to answer the following questions.

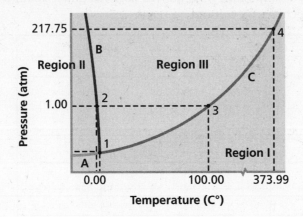

10. What variables are plotted on a phase diagram?

_____

11. What phase of water is represented by each of the following regions?

  a. Region I _____

  b. Region II _____

  c. Region III _____

12. What does point 2 represent?

_____

13. What is the temperature at point 3?

_____

14. What does line A represent?

_____

15. What is point 4 called? What does it represent?

_____

_____

Name _____ Date _____ Class _____

Gases

Section 14.1 The Gas Laws

In your textbook, read about the basic concepts of the three gas laws.

Use each of the terms below to complete the passage. Each term may be used more than once.

pressure	temperature	volume

Boyle's law relates **(1)**_____ and **(2)**_____ if

(3)_____ and amount of gas are held constant. Charles's law relates

(4)_____ and **(5)**_____ if **(6)**_____

and amount of gas are held constant. Gay-Lussac's law relates **(7)**_____

and **(8)**_____ if **(9)**_____ and amount of gas are

held constant.

In your textbook, read about the effects of changing conditions on a sample of gas.

For each question below, write *increases*, *decreases*, or *stays the same*.

_____ **10.** The room temperature increases from 20°C to 24°C. What happens to the pressure inside a cylinder of oxygen contained in the room?

_____ **11.** What happens to the pressure of the gas in an inflated expandable balloon if the temperature is increased?

_____ **12.** An aerosol can of air freshener is sprayed into a room. What happens to the pressure of the gas if its temperature stays constant?

_____ **13.** The volume of air in human lungs increases before it is exhaled. What happens to the temperature of the air in the lungs to cause this change, assuming pressure stays constant?

_____ **14.** A leftover hamburger patty is sealed in a plastic bag and placed in the refrigerator. What happens to the volume of the air in the bag?

_____ **15.** What happens to the pressure of a gas in a lightbulb a few minutes after the light is turned on?

CHAPTER 14 **STUDY GUIDE FOR CONTENT MASTERY**

Section 14.2 The Combined Gas Law and Avogadro's Principle

In your textbook, read about the combined gas law.

Fill in the following table. State what gas law is derived from the combined gas law when the variable listed in the first column stays constant and the variables in the second column change.

Derivations from the Combined Gas Law		
Stays constant	**Change**	**Becomes this law**
Volume	Temperature, pressure	**1.**
Temperature	Pressure, volume	**2.**
Pressure	Temperature, volume	**3.**

In your textbook, read about the relationships among temperature, pressure, and volume of a sample of gas.

Fill in the blanks between the variables in the following concept map to show whether the variables are directly or inversely proportional to each other. Write *direct* or *inverse* between the variables.

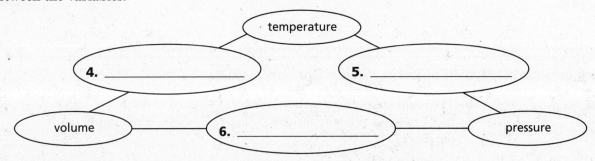

In your textbook, read about the combined gas law and Avogadro's principle.

Circle the letter of the choice that best completes the statement or answers the question.

7. The variable that stays constant when using the combined gas law is

   a. amount of gas.   **b.** pressure.   **c.** temperature.   **d.** volume.

8. The equation for the combined gas law can be used instead of which of the following equations?

   a. Boyle's law   **b.** Charles's law   **c.** Gay-Lussac's law   **d.** all of these

9. Which of the following expresses Avogadro's principle?

   a. Equal volumes of gases at the same temperature and pressure contain equal numbers of particles.

   b. One mole of any gas will occupy a certain volume at STP.

   c. STP stands for standard temperature and pressure.

   d. The molar volume of a gas is the volume that one mole occupies at STP.

CHAPTER  14   **STUDY GUIDE FOR CONTENT MASTERY**

Section 14.2 *continued*

Answer the following questions.

10. What is standard temperature and pressure (STP)?

_____

11. What is the molar volume of a gas equal to at STP?

_____

In your textbook, read about how to solve problems using the combined gas law and Avogadro's principle.

Each problem below needs more information to determine the answer. List as many letters as are needed to solve the problem.

a.	molar volume of the gas	**d.**	pressure of the gas
b.	molar mass of the gas	**e.**	volume of the gas
c.	temperature of the gas	**f.**	No further information is needed.

_____ **12.** What volume will 1.0 g N_2 gas occupy at STP?

_____ **13.** What volume will 2.4 mol He occupy at STP?

_____ **14.** A gas sample occupies 3.7 L at 4.0 atm and 25°C. What volume will the sample occupy at 27°C?

_____ **15.** A sample of carbon dioxide is at 273 K and 244 kPa. What will its volume be at 400 kPa?

_____ **16.** A sample of oxygen occupies 10.0 L at 4.00 atm pressure. At what temperature will the pressure equal 3.00 atm if the final volume is 8.00 L?

_____ **17.** At what pressure will a sample of gas occupy 5.0 L at 25°C if it occupies 3.2 L at 1.3 atm pressure and 20°C?

_____ **18.** How many grams of helium are in a 2-L balloon at STP?

_____ **19.** One mole of hydrogen gas occupies 22.4 L. What volume will the sample occupy if the temperature is 290 K and the pressure is 2.0 atm?

CHAPTER 14   **STUDY GUIDE FOR CONTENT MASTERY**

Section 14.3 **The Ideal Gas Law**

In your textbook, read about the ideal gas law.

Answer the following questions.

1. Why is the mathematical relationship among the amount, volume, temperature, and pressure of a gas sample called the ideal gas law?

_____

_____

2. Define the ideal gas constant, R.

_____

_____

3. In Table 14-1 in your textbook, why does R have different numerical values?

_____

_____

4. What variable is considered in the ideal gas law that is not considered in the combined gas law?

_____

In your textbook, read about real versus ideal gases.

For each statement below, write *true* or *false*.

_____ **5.** An ideal gas is one whose particles take up space.

_____ **6.** At low temperatures, ideal gases liquefy.

_____ **7.** In the real world, gases consisting of small molecules are the only gases that are truly ideal.

_____ **8.** Most gases behave like ideal gases at many temperatures and pressures.

_____ **9.** No intermolecular attractive forces exist in an ideal gas.

_____ **10.** Nonpolar gas molecules behave more like ideal gases than do gas molecules that are polar.

_____ **11.** Real gases deviate most from ideal gas behavior at high pressures and low temperatures.

_____ **12.** The smaller the gas molecule, the more the gas behaves like an ideal gas.

CHAPTER 14

STUDY GUIDE FOR CONTENT MASTERY

Section 14.3 *continued*

In your textbook, read about applying the ideal gas law.

Rearrange the ideal gas law, $PV = nRT$, to solve for each of the following variables. Write your answers in the table.

Rearranging the Ideal Gas Law Equation	
Variable to Find	**Rearranged Ideal Gas Law Equation**
n	**13.**
P	**14.**
T	**15.**
V	**16.**

In your textbook, read about using the ideal gas law to solve for molar mass, mass, or density.

Use the following terms below to complete the statements. Each term may be used more than once.

mass	molar mass	volume

The number of moles of a gas is equal to the **(17)**_____ divided by the

(18)_____.

Density is defined as **(19)**_____ per unit **(20)**_____.

To solve for M in the equation $M = \dfrac{mRT}{PV}$, the **(21)**_____ and the

(22)_____ of the gas must be known.

According to the equation $D = \dfrac{MP}{RT}$, the **(23)**_____ of the gas must be

known when calculating density.

CHAPTER 14   **STUDY GUIDE FOR CONTENT MASTERY**

Section 14.4 Gas Stoichiometry

In your textbook, read about gas stoichiometry.

Balance the following chemical equation. Then use the balanced equation to answer the questions.

1. _____$H_2(g)$ + _____$O_2(g) \rightarrow$ _____$H_2O(g)$

2. List at least two types of information provided by the coefficients in the equation.

_____

_____

3. If 4.0 L of water vapor is produced, what volume of hydrogen reacted? What volume of oxygen?

_____

4. If it is known that 2 mol of hydrogen reacts, what additional information would you need to know to find the volume of oxygen that would react with it?

_____

_____

_____

5. List the steps you would use to find the mass of oxygen that would react with a known number of moles of hydrogen.

_____

_____

_____

6. Find the mass of water produced from 4.00 L H_2 at STP if all of it reacts. Show your work.

CHAPTER 15 STUDY GUIDE FOR CONTENT MASTERY

Solutions

Section 15.1 What are solutions?

In your textbook, read about the characteristics of solutions.

Use each of the terms below just once to complete the passage.

immiscible	liquid	soluble	solution
insoluble	miscible	solute	solvent

Air is a(n) **(1)**_____ of oxygen gas dissolved in nitrogen

gas. The oxygen in air is the **(2)**_____, and nitrogen is the

(3)_____. Because oxygen gas dissolves in a solvent, oxygen gas

is a(n) **(4)**_____ substance. A substance that does not dissolve is

(5)_____. **(6)**_____ solutions are the most common

type of solutions. If one liquid is soluble in another liquid, such as acetic acid in water, the

two liquids are **(7)**_____. However, if one liquid is insoluble in another,

the liquids are **(8)**_____.

Read about solvation in aqueous solutions in your textbook.

The diagram shows the hydration of solid sodium chloride to form an aqueous solution. Use the diagram to answer the following questions.

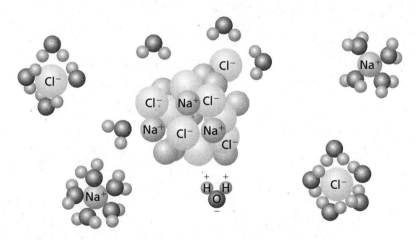

9. Hydration is solvation in which the solvent is water. What is solvation?

_____

_____

Section 15.1 *continued*

10. As sodium chloride dissolves in water, what happens to the sodium and chloride ions?

_____

_____

11. Explain the orientation of the water molecules around the sodium ions and chloride ions.

_____

_____

_____

_____

_____

12. How does the strength of the attraction between water molecules and sodium and chloride ions compare with the strength of the attraction between the sodium ions and chloride ions? How do you know?

_____

_____

_____

_____

13. List three ways that the rate of solvation may be increased.

_____

_____

In your textbook, read about heat of solution, solubility, and factors that affect solubility.

For each statement below, write *true* or *false*.

_____ **14.** The overall energy change that occurs when a solution forms is called the heat of solution.

_____ **15.** Solubility is a measure of the minimum amount of solute that dissolves in a given amount of solvent at a specified temperature and pressure.

_____ **16.** Solvation continues as long as the solvation rate is less than the crystallization rate.

_____ **17.** In a saturated solution, solvation and crystallization are in equilibrium.

_____ **18.** Additional solute can be dissolved in an unsaturated solution.

_____ **19.** The solubility of a gas dissolved in a liquid decreases as the temperature of the solution increases.

Section 15.2 **Solution Concentration**

In your textbook, read about expressing concentration and using percent to describe concentration.

Data related to aqueous solutions of sodium chloride (NaCl) and aqueous solutions of ethanol (C_2H_5OH) are provided in the table below. Use the table to answer the following questions. Circle the letter of the choice that best answers the question.

Solution	Mass (g)		Solution	Volume (mL)	
	NaCl	H_2O		C_2H_5OH	H_2O
1	3.0	100.0	5	2.0	100.0
2	3.0	200.0	6	5.0	100.0
3	3.0	300.0	7	9.0	100.0
4	3.0	400.0	8	15.0	100.0

1. What is the percent by mass of NaCl in solution 1?
   a. 0.030%   **b.** 2.9%   **c.** 3.0%   **d.** 33%

2. Which of the following solutions is the most dilute?
   a. Solution 1   **b.** Solution 2   **c.** Solution 3   **d.** Solution 4

3. What is the percent by volume of C_2H_5OH in Solution 5?
   a. 0.2%   **b.** 1.9%   **c.** 2.0%   **d.** 22%

4. Which of the following solutions is the most concentrated?
   a. Solution 5   **b.** Solution 6   **c.** Solution 7   **d.** Solution 8

In your textbook, read about molarity and preparing molar solutions.

Read the following problem and then answer the questions.

An 85.0-mL aqueous solution contains 7.54 g iron(II) chloride ($FeCl_2$). Calculate the molarity of the solution.

5. What is the mass of the solute? _____

6. What is the volume of the solution? _____

7. Write the equation that is used to calculate molarity.

   _____

8. In what unit must the amount of the solute be expressed to calculate molarity? _____

9. In what unit must the volume of the solution be expressed to calculate molarity? _____

10. Write the expression needed to convert the volume of the solution given in the problem

    to the volume needed to calculate molarity. _____

Section 15.2 *continued*

11. What quantity must be used to convert the mass of the solute given in the problem to the amount of solute needed to calculate molarity?

_____

12. Write the expression used to calculate the amount of solute.

_____

13. Calculate the molarity of the solution. Show all your work.

In your textbook, read about molality and mole fractions.

Answer the following questions.

14. How does molality differ from molarity?

_____

_____

_____

15. Calculate the molality of a solution of 15.4 g sodium bromide (NaBr) dissolved in 125 g of water. Show all your work.

16. What is mole fraction?

_____

_____

17. Calculate the mole fraction of HCl in an aqueous solution that contains 33.6% HCl by mass. Show all your work.

Section 15.3 Colligative Properties of Solutions

In your textbook, read about electrolytes and colligative properties, vapor pressure lowering, boiling point elevation, and freezing point depression.

Use the table to answer the following questions.

Solution	Density (g/L)	Boiling Point (°C)	Freezing Point (°C)
1.0m C$_2$H$_5$OH(aq)	1.05	100.5	−1.8
1.0m HCl(aq)	1.03	101.0	−3.7
1.0m NaCl(aq)	1.06	101.0	−3.7
2.0m NaCl(aq)	1.12	102.1	−7.4

1. Which properties in the table are colligative properties?

_____

2. What can you conclude about the relationship between colligative properties and the number of ions in solution from the 1.0m NaCl(aq) and 2.0m NaCl(aq) solutions?

_____

3. What can you conclude about the relationship between colligative properties and the type of ions in solution from the 1.0m HCl(aq) and 1.0m NaCl(aq) solutions?

_____

Suppose that in a simple system, a semipermeable membrane is used to separate a sucrose-water solution from its pure solvent, water. Match the descriptions of the system in Column A with the terms in Column B.

Column A	Column B
_____ **4.** Cannot cross the semipermeable membrane	**a.** osmotic pressure
_____ **5.** Can cross the semipermeable membrane	**b.** water molecules
_____ **6.** The side that exerts osmotic pressure	**c.** semipermeable membrane
_____ **7.** The diffusion of the solvent particles across the semipermeable membrane from the area of higher solvent concentration to the area of lower solvent concentration	**d.** sugar molecules
	e. osmosis
_____ **8.** The barrier with tiny pores that allow some particles to pass through but not others	**f.** solution side
	g. pure solvent side
_____ **9.** The side from which more water molecules cross the semipermeable membrane	
_____ **10.** A colligative property of solutions	

Section 15.4 **Heterogeneous Mixtures**

In your textbook, read about suspensions and colloids.

For each statement below, write *true* or *false*.

_____ 1. A solution is a mixture containing particles that settle out of the mixture if left undisturbed.

_____ 2. The most abundant substance in a colloid is the dispersion medium.

_____ 3. A colloid can be separated by filtration.

_____ 4. A solid emulsion consists of a liquid dispersed in a solid.

_____ 5. Whipped cream is an example of a foam.

_____ 6. In an aerosol, the dispersing medium is a liquid.

_____ 7. Brownian motion results from the collisions of particles of the dispersion medium with the dispersed particles.

_____ 8. Dispersed particles in a colloid do not tend to settle out because they have polar or charged atomic groups on their surfaces.

_____ 9. Stirring an electrolyte into a colloid stabilizes the colloid.

_____ 10. Colloids demonstrate the Tyndall effect.

The table below lists the characteristics of particles in colloids, solutions, and suspensions. Place a check in the column of each mixture whose particles have a particular characteristic.

Characteristics of Particles	Colloid	Solution	Suspension
11. Less than 1 nm in diameter			
12. Between 1 nm and 1000 nm in diameter			
13. More than 1000 nm in diameter			
14. Settle out if undisturbed			
15. Pass through standard filter paper			
16. Lower vapor pressure			
17. Scatter light			

Name _____ Date _____ Class _____

Energy and Chemical Change

Section 16.1 Energy
In your textbook, read about the nature of energy.

In the space at the left, write *true* if the statement is true; if the statement is false, change the italicized word or phrase to make it true.

_____ **1.** *Energy* is the ability to do work or produce heat.

_____ **2.** The law of conservation of energy states that energy *can be* created and destroyed.

_____ **3.** Chemical potential energy is energy stored in a substance because of its *composition*.

_____ **4.** *Heat* is a form of energy that flows from a warmer object to a cooler object.

_____ **5.** A calorie is the amount of energy required to raise the temperature of *one gram* of pure water by one degree Celsius.

_____ **6.** A *calorie* is the SI unit of heat and energy.

_____ **7.** The *specific heat* of a substance is the amount of heat required to raise the temperature of one gram of that substance by one degree Celsius.

_____ **8.** *Kinetic energy* is energy of motion.

_____ **9.** Chemicals participating in a chemical reaction contain *only potential energy*.

_____ **10.** One nutritional Calorie is equal to *100 calories*.

_____ **11.** One calorie equals *4.184 joules*.

_____ **12.** When a fuel is burned, some of its *chemical potential energy* is lost as heat.

_____ **13.** To convert kilojoules to joules, *divide* the number of kilojoules by 1000 joules/1 kilojoule.

Answer the following question. Show all your work.

14. If the temperature of a 500.0-g sample of liquid water is raised 2.00°C, how much heat is absorbed by the water? The specific heat of liquid water is 4.184 J/(g·°C).

CHAPTER 16   STUDY GUIDE FOR CONTENT MASTERY

Section 16.2 Heat in Chemical Reactions and Processes

In your textbook, read about measuring heat and about chemical energy and the universe.

For each item in Column A, write the letter of the matching item in Column B.

<table>
<tr><td colspan="2" align="center">Column A</td><td align="center">Column B</td></tr>
<tr><td>_____</td><td>1. An insulated device used to measure the amount of heat absorbed or released during a chemical or physical process</td><td>a. system</td></tr>
<tr><td>_____</td><td>2. The study of heat changes that accompany chemical reactions and phase changes</td><td>b. calorimeter

c. thermochemistry</td></tr>
<tr><td>_____</td><td>3. The specific part of the universe that contains the reaction or process you wish to study</td><td>d. universe

e. enthalpy</td></tr>
<tr><td>_____</td><td>4. The change in enthalpy in a chemical reaction</td><td>f. enthalpy (heat) of reaction</td></tr>
<tr><td>_____</td><td>5. A system plus its surroundings</td><td>g. surroundings</td></tr>
<tr><td>_____</td><td>6. The heat content of a system at constant pressure</td><td></td></tr>
<tr><td>_____</td><td>7. Everything in the universe except the system being studied</td><td></td></tr>
</table>

Solution of
$Ba(OH)_2$
and
NH_4NO_3

Use the illustration to answer the following questions.

8. A scientist is studying the solution in the flask. What is the system?

_____

9. What are the surroundings?

_____

10. What is the universe?

_____

CHAPTER 16

STUDY GUIDE FOR CONTENT MASTERY

Section 16.3 Thermochemical Equations

In your textbook, read about writing thermochemical equations and about changes of state.

Use the following terms to complete the statements. Some terms will be used more than once.

thermochemical equation	enthalpy of combustion	released
molar enthalpy of vaporization	molar enthalpy of fusion	absorbs
cool	heat	

1. A(n) _____ is a balanced chemical equation that includes the physical states of all reactants and products and the energy change that accompanies the reaction.

2. The enthalpy change for the complete burning of one mole of a substance is the

   _____.

3. The _____ is the heat required to vaporize one mole of a liquid.

4. The _____ is the heat required to melt one mole of a solid substance.

5. Converting two moles of a liquid to a solid requires an amount of energy that is twice

   the _____.

6. $2H_2(g) + O_2(g) \rightarrow 2H_2O(g)$ $\Delta H = -572$ kJ is a(n) _____.

7. The conversion of a gas to a liquid involves the _____.

8. When a gas condenses to a liquid, heat is _____ to the surroundings.

9. Sweating makes you feel cooler because, as it evaporates, the water on your skin

   _____ heat from your body.

10. If you put an ice cube in a glass of soda pop, the heat absorbed by the ice will cause the

    ice to melt, and the soda pop will become _____.

11. If it takes 100 joules to melt a piece of ice, _____ must be absorbed by the ice.

12. In the equation $H_2O(s) \rightarrow H_2O(l)$ $\Delta H = 600$ kJ, the positive value for ΔH means that

    _____ is absorbed in the reaction.

CHAPTER 16

STUDY GUIDE FOR CONTENT MASTERY

Section 16.4 Calculating Enthalpy Change

In your textbook, read about Hess's law and standard enthalpy (heat) of formation.

In the space at the left, write *true* if the statement is true; if the statement is false, change the italicized word or phrase to make it true.

_____ 1. Hess's law states that if two or more thermochemical equations can be added to produce a final equation for a reaction, then the sum of all the enthalpy changes for the individual reactions is the enthalpy change for the *final reaction*.

_____ 2. The standard enthalpy of formation is the change in enthalpy that accompanies the formation of *one gram* of a compound in its standard state from its constituent elements in their standard states.

_____ 3. The standard state of iron is *solid*.

_____ 4. For a pure gas, the standard state is the gas at a pressure of *one atmosphere*.

_____ 5. The symbol used to represent standard enthalpy of formation is ΔH_f°.

_____ 6. The standard state of a substance is the normal state of the substance at *0 K* and one atmosphere pressure.

_____ 7. The standard enthalpy of formation of a free element in its standard state is *0.0 kJ*.

_____ 8. A standard enthalpy of formation that has a *negative* value means that energy is absorbed during the reaction.

_____ 9. The standard state of oxygen is *gas*.

_____ 10. Standard enthalpies of formation provide data for calculating the enthalpies of reactions under standard conditions using *Hess's law*.

_____ 11. The standard state of mercury is *solid*.

Section 16.4 *continued*

Use the table to answer the following questions.

Compound	Formation Equation	$\Delta H_f°$ (kJ/mol)
$CH_4(g)$	$C(graphite) + 2H_2(g) \longrightarrow CH_4(g)$	75
$CH_3OH(g)$	$C(graphite) + 2H_2(g) + \frac{1}{2}O_2(g) \longrightarrow CH_3OH(g)$	239
$H_2O(g)$	$\frac{1}{2}O_2(g) + H_2(g) \longrightarrow H_2O(g)$	242

12. What does a formation equation show?

____

____

____

____

13. What does the negative sign on the value of an enthalpy of formation indicate?

____

____

14. Using the formation equations for $CH_4(g)$, $CH_3OH(g)$, and $H_2O(g)$, calculate ΔH_{rxn} for the following equation. Show and explain all your work.

$$CH_4(g) + H_2O(g) \longrightarrow CH_3OH(g) + H_2(g)$$

CHAPTER 16   **STUDY GUIDE FOR CONTENT MASTERY**

Section 16.5 **Reaction Spontaneity**

In your textbook, read about spontaneous processes and about entropy, the universe, and free energy.

Use each of the terms below to complete the statements.

spontaneous process	entropy	law of disorder	free energy

1. A(n) _____ is a physical or chemical change that occurs with no

   outside intervention.

2. A measure of disorder or randomness of the particles that make up a system is called

   _____ .

3. The _____ states that spontaneous processes always proceed in such

   a way that the entropy of the universe increases.

4. _____ is the energy that is available to do work.

For each statement below, write *true* or *false*.

_____   5. A process cannot be spontaneous if it is exothermic and there is an increase in disorder.

_____   6. A process cannot be spontaneous if it is endothermic and there is a decrease in disorder.

_____   7. A process cannot be spontaneous if it is exothermic and there is a decrease in disorder as long as the temperature remains low.

_____   8. A process cannot be spontaneous if it is endothermic and there is an increase in disorder as long as the temperature remains high.

_____   9. A process can never be spontaneous if the entropy of the universe increases.

_____   10. When ΔG for a reaction is negative, the reaction is spontaneous.

_____   11. When ΔG for a reaction is positive, the reaction is not spontaneous.

_____   12. When ΔH for a reaction is negative, the reaction is never spontaneous.

_____   13. When ΔH for a reaction is large and positive, the reaction is not expected to be spontaneous.

Name _____ Date _____ Class _____

Reaction Rates

Section 17.1 A Model for Reaction Rates

In your textbook, read about expressing reaction rates and explaining reactions and their rates.

Use each of the terms below just once to complete the passage.

collision theory	activated complex	transition state
activation energy	reaction rate	mol/(L·s)

According to the **(1)**_____, atoms, ions, and molecules must collide in

order to react. Once formed, the **(2)**_____ is a temporary, unstable

arrangement of atoms that may then form products or may break apart to reform the reactants.

This physical arrangement is known as the **(3)**_____. Every chemical

reaction requires energy, and the minimum amount of energy that reacting particles must have

to form the activated complex is the **(4)**_____. In a chemical reaction, the

(5)_____ is the change in concentration of a reactant or product per unit

time. It may be expressed using the units of **(6)**_____.

Use the energy diagram for the rearrangement reaction of methyl isonitrile to acetonitrile to answer the following questions.

7. What kind of reaction is represented by this diagram, endothermic or exothermic?

_____

8. What is the chemical structure identified at the top of the curve on the diagram?

_____

9. What does the symbol E_a represent?

_____

10. What does the symbol ΔE represent?

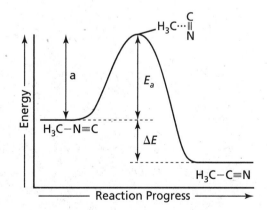

_____

Section 17.1 *continued*

For each item in Column A, write the letter of the matching item in Column B.

Column A	Column B
_____ **11.** Expresses the average rate of loss of a reactant	**a.** average reaction rate
_____ **12.** Expressed as Δquantity/Δtime	**b.** positive number
_____ **13.** Expresses the average rate of formation of a product	**c.** negative number

Use the figure below to answer the following questions.

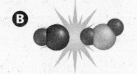

    → ←   ← →   → ←
   Collision   Rebound   Collision   Activated complex   Products
   Incorrect orientation   **Correct orientation**

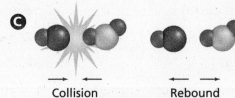

   → ←   ← →
  Collision   Rebound
 Correct orientation
 Insufficient energy

14. What molecules collided in collisions A, B, and C? _____

15. What do the arrows represent?

_____

_____

16. Which collision(s) formed products? What were the products? _____

17. Explain why the other collision(s) did not form products.

_____

_____

_____

_____

18. Which collision(s) formed an activated complex? Identify the activated complex.

_____

CHAPTER 17 STUDY GUIDE FOR CONTENT MASTERY

Section 17.2 Factors Affecting Reaction Rates

In your textbook, read about the factors that affect reaction rates (reactivity, concentration, surface, area, temperature, and catalysts).

In the space at the left, write *true* if the statement is true; if the statement is false, change the italicized word to make it true.

_____ **1.** *Decreasing* the concentration of reactants increases the collision frequency between reacting particles.

_____ **2.** A *heterogeneous* catalyst exists in a different physical state than the reaction it catalyzes.

_____ **3.** Increasing the *concentration* of a substance increases the kinetic energy of the particles that make up the substance.

_____ **4.** Catalysts increase the rates of chemical reactions by *raising* the activation energy of the reactions.

_____ **5.** *Increasing* the surface area of a reactant increases the rate of the reaction.

_____ **6.** Raising the temperature of a reaction increases the rate of the reaction by increasing the *energy* of the collisions between reacting particles.

Answer the following questions.

7. A chemist heated a sample of steel wool in a burner flame exposed to oxygen in the air. He also heated a sample of steel wool in a container of nearly 100% oxygen. The steel-wool sample in the container reacted faster than the other sample. Explain why.

_____

_____

8. Would the chemist have observed the same results if he used a block of steel instead of steel wool? Explain your answer.

_____

_____

9. How would the reaction have differed if the steel wool was not heated?

_____

_____

CHAPTER 17   **STUDY GUIDE FOR CONTENT MASTERY**

Section 17.3 **Reaction Rate Laws**

In your textbook, read about reaction rate laws and determining reaction order.

Use each of the terms below to complete the statements.

chemical reaction	rate law	specific rate constant
reaction orders	concentration	time

Equation 1 $aA + bB \rightarrow cC + dD$

Equation 2 $-\dfrac{\Delta[A]}{\Delta t} = k[A]^m[B]^n$

1. Equation 1 describes a _____.

2. Equation 2 expresses the mathematical relationship between the rate of a chemical

reaction and the concentrations of the reactants. This is known as the

_____.

3. The variable k in equation 2 is the _____, a numerical value that

relates the reaction rate and the concentration at a given temperature.

4. The variables m and n are the _____. These define how the rate is

affected by the concentrations of the reactants.

5. The square brackets [ ] represent _____.

6. The variable t represents _____.

Answer the questions about the following rate law.

Rate $= k[A]^1[B]^2$

7. What is the reaction order with respect to A? _____

8. What is the reaction order with respect to B? _____

9. What is the overall reaction order for the rate law? _____

10. Doubling the concentration of A will cause the rate to double. What would happen if you
doubled the concentration of B?

_____

11. A reaction rate can be expressed as Rate $= k[A]^2$. What is the reaction order for this reaction?

_____

Section 17.4 Instantaneous Reaction Rates and Reaction Mechanisms

In your textbook, read about instantaneous reaction rates.

Circle the letter of the choice that best completes the statement.

1. _____ is determined by finding the slope of the straight line tangent to the curve of a plot of the change in concentration of a reactant versus time.

   a. Instantaneous rate   **c.** Reaction mechanism

   b. Change in temperature   **d.** Reaction order

2. A(n) _____ consists of two or more elementary steps.

   a. complex reaction   **c.** reaction mechanism

   b. elementary step   **d.** reaction order

3. A(n)_____ is a substance produced in an elementary step and consumed in another elementary step.

   a. instantaneous rate   **c.** reaction mechanism

   b. intermediate   **d.** rate-determining step

4. A(n) _____ is the complete sequence of elementary reactions that make up a complex reaction.

   a. instantaneous rate   **c.** reaction mechanism

   b. elementary step   **d.** reaction order

5. The _____ is the slowest of the elementary steps in a complex reaction.

   a. instantaneous rate   **c.** rate-determining step

   b. intermediate   **d.** reaction order

6. The _____ can be used to determine the instantaneous rate for a chemical reaction.

   a. rate-determining step   **c.** products

   b. intermediates   **d.** rate law

7. An element or compound that reacts in one step of a complex reaction and reforms in a another step of the complex reaction is

   a. an intermediate.

   b. a catalyst.

   c. not part of the reaction mechanism.

   d. shown in the net chemical equation for the reaction.

Section 17.4 *continued*

In the space at the left, write *true* if the statement is true; if the statement is false, change the italicized word or phrase to make it true.

_____ **8.** To determine the *instantaneous rate*, you must know the specific rate constant, the concentrations of the reactants, and the reaction orders for the reaction.

_____ **9.** A reaction rate that is defined as k[A][B] and that has a specific rate constant of 1.0×10^1 L/(mol·s), [A] = 0.1M, and [B] = 0.1M would have an instantaneous rate of *0.01 mol/(L·s)*.

In your textbook, read about reaction mechanisms.

Answer the following questions about the proposed reaction mechanism for the complex reaction below.

$$2NO(g) + 2H_2(g) \rightarrow N_2(g) + 2H_2O(g)$$

Proposed Mechanism

$$2NO \rightarrow N_2O_2 \qquad \text{(fast)}$$

$$N_2O_2 + H_2 \rightarrow N_2O + H_2O \qquad \text{(slow)}$$

$$N_2O + H_2 \rightarrow N_2 + H_2O \qquad \text{(fast)}$$

10. How many elementary steps make up the complex reaction?

_____

11. What is the rate-determining step for this reaction?

_____

12. What are N_2O_2 and N_2O in the reaction?

_____

13. Is there a catalyst involved in the reaction? Explain your answer.

_____

_____

14. What can you conclude about the activation energy for the rate-determining step?

_____

15. If you wanted to increase the rate of the overall reaction, what would you do?

_____

CHAPTER 18 **STUDY GUIDE FOR CONTENT MASTERY**

Chemical Equilibrium

Section 18.1 Equilibrium: A State of Dynamic Balance

In your textbook, read about chemical equilibrium.

Complete each statement.

1. When a reaction results in almost complete conversion of reactants to products, chemists

   say the reaction goes to _____.

2. A reaction that can occur in both the forward and the reverse directions is called a(n)

   _____.

3. _____ is a state in which the forward and reverse reactions balance

   each other because they take place at equal rates.

4. At equilibrium, the concentrations of reactants and products are _____,

   but that does not mean that the amounts or concentrations are _____.

5. Equilibrium is a state of _____, not one of _____.

In your textbook, read about equilibrium expressions and constants.

For each statement below, write *true* or *false*.

_____ 6. The law of chemical equilibrium states that at a given pressure, a chemical system may reach a state in which a particular ratio of reactant to product concentrations has a constant value.

_____ 7. The equation $H_2(g) + I_2(g) \rightleftharpoons 2HI(g)$ is an example of a homogeneous equilibrium.

_____ 8. If an equilibrium constant has a value less than one, the reactants are favored at equilibrium.

_____ 9. The value for K_{eq} is constant only at a specific volume.

_____ 10. If the equilibrium constant for a reaction at 300 K is 49.7, the concentration of the reactants will be greater than the concentration of the products.

_____ 11. A heterogeneous equilibrium means that reactants and products are present in more than one state.

_____ 12. The product of the forward chemical reaction is HI, for the equilibrium expression:

$$K_{eq} = \frac{[HI]^2}{[H_2][I_2]}$$

Section 18.1 *continued*

In your textbook, read about determining equilibrium constants.

A chemist did two experiments to determine the equilibrium constant for the reaction of sulfur dioxide with oxygen to form sulfur trioxide. Use the table showing the results of the experiments to answer the following questions.

$2SO_2(g) + O_2(g) \rightleftharpoons 2SO_3(g)$ at 873 K			
Experiment 1		**Experiment 2**	
Initial concentrations	Equilibrium concentrations	Initial concentrations	Equilibrium concentration
$[SO_2] = 2.00M$	$[SO_2] = 1.50M$	$[SO_2] = 0.500M$	$[SO_2] = 0.590M$
$[O_2] = 1.50M$	$[O_2] = 1.26M$	$[O_2] = 0M$	$[O_2] = 0.0450M$
$[SO_3] = 3.00M$	$[SO_3] = 3.50M$	$[SO_3] = 0.350M$	$[SO_3] = 0.260M$

13. Write the equation to calculate the equilibrium constant for the reaction.

_____

14. Is this reaction an example of a homogeneous or heterogeneous equilibrium?

_____

15. Calculate the equilibrium constant from the data obtained in experiment 1.

_____

_____

16. What is the equilibrium constant for the reaction in experiment 2?

_____

_____

17. Was it necessary to calculate both equilibrium constants? Why or why not?

_____

_____

18. What does this experiment show about the initial concentrations of products and reactants in a reversible reaction?

_____

_____

_____

Section 18.2 Factors Affecting Chemical Equilibrium

In your textbook, read about Le Châtelier's Principle.

Answer the following questions.

1. What does Le Châtelier's Principle say?

_____

_____

2. What are three kinds of stresses that can be placed on a system?

_____

_____

For each reaction below, state the direction, left or right, in which the equilibrium will shift when the indicated substance is added. Identify one other way in which the reaction could be shifted in the same direction you indicated. (Hint: There may be more than one way to do this.)

3. Reaction: $N_2(g) + 3H_2(g) \rightleftharpoons 2NH_3(g)$; NH_3 added

_____

4. Reaction: $H_2(g) + I_2(g) \rightleftharpoons 2HI(g)$; H_2 added

_____

5. Reaction: $CO(g) + H_2O \rightleftharpoons CO_2(g) + H_2(g)$; H_2O added

_____

6. Reaction: $2SO_2(g) + O_2(g) \rightleftharpoons 2SO_3(g)$; SO_3 added

_____

7. Reaction: $2SO_2(g) + O_2(g) \rightleftharpoons 2SO_3(g)$; SO_2 added

_____

8. Reaction: $2NCl_3(g) \rightleftharpoons N_2(g) + 3Cl_2(g)$; NCl_3 added

_____

Section 18.2 *continued*

In your textbook, read about factors affecting chemical equilibrium.

Use each of the terms below just once to complete the passage.

right	exothermic	increase	stress	catalyst	energy
smallest	change	reverse	constant	forward	

When you decrease the volume of a reaction vessel, you **(9)**_____

the pressure. This causes a reaction at equilibrium to shift to the side with the

(10)_____ number of moles. If the reaction has an equal number of

moles of reactants and products, changing the volume of the reaction vessel causes no

(11)_____ in the equilibrium.

Changing the temperature of a reaction at equilibrium alters both the equilibrium

(12)_____ and the equilibrium position. When a reaction is

(13)_____, which means it releases energy, lowering the temperature

shifts the equilibrium to the **(14)**_____ because the forward reaction

liberates heat and removes the **(15)**_____.

A **(16)**_____ speeds up a reaction by lowering the

(17)_____ requirements for the reaction, but it does so equally in both the

(18)_____ and the **(19)**_____ directions. The reaction

will reach equilibrium more quickly, but with no change in the amount of product formed.

For each reaction below, indicate in which direction the equilibrium shifts when the stated stress is applied to the system. Write *R* if the reaction shifts to the right, *L* if it shifts to the left, or *NC* if there is no change.

	Reaction	**Stress**
_____	**20.** $PCl_5(g) \rightleftharpoons PCl_3(g) + Cl_2(g)$ + heat	temperature increase
_____	**21.** $CO(g) + Fe_3O_4(s) \rightleftharpoons CO_2(g) + 3FeO(s)$	volume increase
_____	**22.** $C_2H_2(g) + H_2O(g) \rightleftharpoons CH_3CHO(g)$ + heat	temperature decrease
_____	**23.** $2NO(g) + H_2(g) \rightleftharpoons N_2O(g) + H_2O(g)$ + heat	volume decrease
_____	**24.** Heat + $H_2(g) + I_2(g) \rightleftharpoons 2HI(g)$	temperature decrease
_____	**25.** $H_2(g) + Cl_2(g) \rightleftharpoons 2HCl(g)$ + heat	volume decrease

CHAPTER 18

Section 18.3 **Using Equilibrium Constants**

In your textbook, read about calculating equilibrium concentrations.

Answer the following questions.

1. What can you use the equilibrium constant to do?

   _____

   _____

2. Given the reaction: $N_2 + O_2 \rightleftharpoons 2NO$ for which the K_{eq} at 2273 K is 1.2×10^{-4}

   a. Write the equilibrium constant expression for the reaction.

   _____

   b. Write the equation that would allow you solve for the concentration of NO.

   _____

   c. What is the concentration of NO if $[N_2] = 0.166M$ and $[O_2] = 0.145M$?

   _____

   _____

   _____

3. What is the solubility product constant?

   _____

   _____

4. What is the solubility product constant expression for the reaction:
   $Mg_3(PO_4)_2(s) \rightleftharpoons 3Mg^{2+}(aq) + 2PO_4^{3-}(aq)$

   _____

5. Given the equilibrium $BaSO_4(s) \rightleftharpoons Ba^{2+}(aq) + SO_4^{2-}(aq)$, what is the solubility product constant expression?

   _____

6. The solubility product constant for $BaSO_4$ at 298 K is 1.1×10^{-10}. Calculate the solubility of $BaSO_4$ in mol/L at 298 K.

   _____

   _____

   _____

   _____

Section 18.3 *continued*

In your textbook, read about predicting precipitates.

The solubility product constant can be used to determine if a precipitate will form when two aqueous solutions are mixed together. First, calculate the concentrations of the ions in the final solution. Use the solubility product constant expression to calculate the ion product (Q_{sp}) for the substance that might precipitate. Compare the result with the K_{sp} of the substance.

7. What can you say about a solution when

    a. Q_{sp} is greater than K_{sp}?

_____

_____

    b. Q_{sp} is equal to K_{sp}?

_____

_____

    c. Q_{sp} is less than K_{sp}?

_____

_____

8. Predict whether a precipitate of AgBr will form if 100 mL of 0.0025M AgNO$_3$ and 100 mL of 0.0020M NaBr are mixed.

_____

_____

_____

_____

_____

9. Explain briefly why Ag$_3$PO$_4$ might be more soluble in water than in the same volume of a solution containing Na$_3$PO$_4$.

_____

_____

_____

_____

_____

CHAPTER 19 STUDY GUIDE FOR CONTENT MASTERY

Acids and Bases

Section 19.1 Acids and Bases: An Introduction

In your textbook, read about the properties of acids and bases.

For each description below, write *acid* if it tells about a property of an acid or *base* if it tells about a property of a base. If the property does not apply to either an acid or a base, write *neither*. If it applies to both an acid and a base, write *both*.

_____ **1.** Can turn litmus paper a different color

_____ **2.** Reacts with certain metals

_____ **3.** Contains more hydrogen ions than hydroxide ions

_____ **4.** Feels slippery

_____ **5.** Reacts with carbonates

_____ **6.** Feels rough

_____ **7.** Contains equal numbers of hydrogen and hydroxide ions

_____ **8.** Tastes bitter

_____ **9.** Tastes sour

In your textbook, read about the different models of acids and bases.

Use the terms below to complete the passage. You may use each term more than once.

Arrhenius	Brønsted-Lowry	conjugate acid
conjugate base	hydrogen	hydroxide

The **(10)**_____ model of acids and bases states that an acid contains

the element **(11)**_____ and forms ions of this element when it is dissolved

in water. A base contains the **(12)**_____ group and dissociates to produce

(13)_____ ions in aqueous solution.

According to the **(14)**_____ model, an acid donates

(15)_____ ions, and a base accepts **(16)**_____ ions.

According to this model, in an acid-base reaction, each acid has a

(17)_____, and each base has a **(18)**_____.

Section 19.2 Strengths of Acids and Bases

In your textbook, read about strengths of acids.

Circle the letter of the choice that best completes the statement or answers the question.

1. Acid A and acid B are of equal concentration and are tested with a conductivity apparatus. When the electrodes are placed in acid A, the bulb glows dimly. When they are placed in acid B, the bulb glows more brightly. Which of the following is true?

    a. Acid A is stronger than acid B.

    b. Acid B is stronger than acid A.

    c. Acid A and acid B are of equal strength.

    d. No comparison of strength can be made from the results.

2. A chemical equation for the ionization of an acid uses a single arrow to the right ($\rightarrow$) to separate the reactant and product sides of the equation. Which of the following is true?

    a. The arrow does not indicate relative strength.   **c.** The ionizing acid is strong.

    b. The ionizing acid is half ionized.   **d.** The ionizing acid is weak.

3. Sulfuric acid is a strong acid. What is true about its conjugate base?

    a. Its conjugate base is amphoteric.

    b. Its conjugate base is strong.

    c. Its conjugate base is weak.

    d. No conclusion can be made regarding the strength of the conjugate base.

4. In solution, a weak acid produces

    a. a mixture of molecules and ions.   **c.** all molecules.

    b. all ions.   **d.** anions, but no hydronium ions.

5. Why are K_a values all small numbers?

    a. The concentration of water does not affect the ionization.

    b. The equilibrium is not stable.

    c. The solutions contain a high concentration of ions.

    d. The solutions contain a high concentration of un-ionized acid molecules.

6. Which of the following dissociates entirely into metal ions and hydroxide ions in solution?

    a. a strong acid   **b.** a strong base   **c.** a weak acid   **d.** a weak base

7. In general, compounds formed from active metals, and hydroxide ions are

    a. strong acids.   **b.** strong bases.   **c.** weak acids.   **d.** weak bases.

CHAPTER 19 **STUDY GUIDE FOR CONTENT MASTERY**

Section 19.3 What is pH?

In your textbook, read about the ion product constant for water.

Answer the following questions.

  1. Write the simplest form of the chemical equation for the self-ionization of water.

  _____

  2. Write the equilibrium constant expression, K_{eq}, for this equation.

  _____

  3. Write the expression for the equilibrium constant for water, K_w.

  _____

  4. Why can the concentration of water be ignored in the equilibrium expression for water?

  _____

  5. What is the numerical value of K_w at 298 K?

  _____

  6. In solution, if the hydroxide ion concentration increases, what happens to the hydrogen
     ion concentration?

  _____

  7. If the concentration of hydroxide ions in solution is 1.0×10^{-6}, what is the hydrogen ion
     concentration?

  _____

  8. Is the solution in question 7 acidic, basic, or neutral? Explain.

  _____

In your textbook, read about pH and pOH.

**In the space at the left, write *true* if the statement is true; if the statement is false,
change the italicized word or number to make it true.**

_____ **9.** The pH of a solution is the negative logarithm of its *hydroxide* ion
                              concentration.

_____ **10.** Values for pH range from *0 to 14*.

_____ **11.** Stomach contents can have a pH of 2, which means that they are
                              basic.

_____ **12.** The hydrogen ion concentration in a solution with a pH of 3 is
                              two times greater than the hydrogen ion concentration in a
                              solution with a pH of 5.

Section 19.3 *continued*

_____ **13.** The pH of a neutral solution at room temperature *equals* the pOH
of the solution.

_____ **14.** If the pH of a solution is 3, its pOH is *10*.

_____ **15.** The pH of a solution with a $[H^+]$ of 1×10^{-8} is *8*.

_____ **16.** The pH of a solution with a $[OH^-]$ of 1×10^{-6} is *6*.

In your textbook, read about calculating the pH of acids and bases.

Solve each of the following problems. Show your work.

17. What is the pH of a $4.3 \times 10^{-2} M$ HCl solution? HCl is a strong acid.

18. Calculate the pH of a $5.2 \times 10^{-3} M$ H_2SO_4 solution? H_2SO_4 is a strong acid.

19. What is the pH of a $2.5 \times 10^{-5} M$ NaOH solution? NaOH is a strong base.

20. Calculate the pH of a $3.6 \times 10^{-6} M$ $Ca(OH)_2$ solution. $Ca(OH)_2$ is a strong base.

In your textbook, read about measuring pH.

Complete the passage.

Indicator paper can be used to measure the **(21)** _____ of a solution. Indicators

are substances that are different **(22)** _____ depending on the pH of the solution

tested. Another way to measure the acidity of the solution is the **(23)** _____, which

uses electrodes placed in solution to directly read the results.

CHAPTER 19   **STUDY GUIDE FOR CONTENT MASTERY**

Section 19.4 **Neutralization**

In your textbook, read about neutralization and titration.

For each item in Column A, write the letter of the matching item in Column B.

Column A	Column B
_____ **1.** A chemical dye that changes color based on the pH of a solution	**a.** acid-base indicator
_____ **2.** A method for using a neutralization reaction to determine the concentration of a solution	**b.** end point
	c. equivalence point
_____ **3.** A reaction in which an acid and a base react to produce a salt and water	**d.** neutralization
	e. salt
_____ **4.** A solution of known concentration	**f.** standard solution
_____ **5.** An ionic product of an acid-base reaction	**g.** titration
_____ **6.** The point in a titration in which an indicator changes color	
_____ **7.** The stoichiometric point of a titration	

Complete the following table, indicating the formula and name of the salt formed by a neutralization reaction between the listed acid and base.

Acid	Base	Salt formula	Salt name
8. HCl	KOH	KCl	potassium chloride
9. H_2SO_4	$Mg(OH)_2$		
10. H_3PO_4	$NaOH$		
11. HNO_3	$Fe(OH)_3$		
12. H_3PO_4	$Ca(OH)_2$		

In the space at the left, write *1* through *4* to show the correct sequence of the steps in performing a titration using a pH meter. Then, write *5* through *8* to sequence the steps used to calculate the concentration of the unknown solution.

Sequence of Steps

_____ **13.** Continue adding the standard solution to the solution of unknown concentration until the equivalence point is reached.

_____ **14.** Fill a buret with the standard solution.

_____ **15.** Start adding the standard solution slowly, with mixing, to the solution of unknown concentration, reading the pH at regular intervals.

Section 19.4 *continued*

_____ **16.** Use a pH meter to check the pH of a solution of known volume but unknown concentration.

Calculation

_____ **17.** Calculate the number of moles of acid or base in the volume of standard solution added.

_____ **18.** Use the mole ratio from the balanced equation to calculate the number of moles of reactant in the unknown solution.

_____ **19.** Use the number of moles and volume of the unknown solution to calculate molarity.

_____ **20.** Write the balanced chemical equation for the neutralization reaction.

In your textbook, read about salt hydrolysis.

Complete the following concept map, using the terms *acidic*, *basic*, and *neutral*.

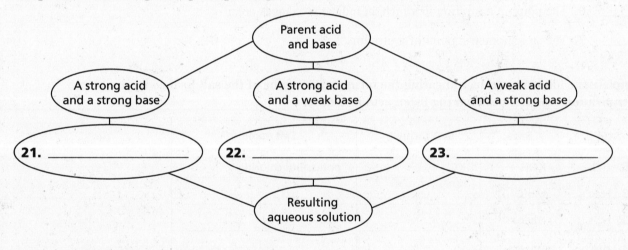

In your textbook, read about buffer solutions.

For each statement below, write *true* or *false*.

_____ **24.** Buffers resist change in pH.

_____ **25.** A buffer can be a mixture of a weak acid and its conjugate base.

_____ **26.** An example of a buffer solution is a mixture of acetic acid and sodium acetate.

_____ **27.** A buffer solution changes pH only a small amount even if large amounts of acid or base are added.

_____ **28.** A buffer system should contain considerably more acid than base.

_____ **29.** Specific buffer systems should be chosen based on the pH that must be maintained.

CHAPTER 20 STUDY GUIDE FOR CONTENT MASTERY

Redox Reactions

Section 20.1 Oxidation and Reduction

In your textbook, read about redox reactions.

Circle the letter of the choice that best completes the statement or answers the question.

1. Redox reactions are characterized by

   a. formation of a solid, a gas, or water.

   b. replacement of one element in a compound by another element.

   c. sharing of electrons.

   d. transfer of electrons.

2. If a calcium atom loses two electrons, it becomes

   a. a Ca^{2-} ion. **b.** an oxidizing agent. **c.** oxidized. **d.** reduced.

3. In a redox reaction, an oxidizing agent is

   a. balanced. **c.** oxidized.

   b. increased in oxidation number. **d.** reduced.

4. An oxidation reaction occurs

   a. at the same time a reduction reaction occurs.

   b. before its corresponding reduction reaction occurs.

   c. independently of any reduction reaction.

   d. only when electrons are gained.

5. Consider the equation $Ca(s) + O_2(g) \rightarrow 2CaO(s)$.
   In this reaction, calcium is oxidized because it

   a. becomes part of a compound. **c.** loses electrons.

   b. gains electrons. **d.** reacts with oxygen.

6. The number of electrons lost by an element when it forms ions is the element's

   a. charge. **b.** oxidation number. **c.** reduction number. **d.** shared electrons.

7. A loss of electrons is

   a. oxidation. **b.** oxidation–reduction. **c.** redox. **d.** reduction.

8. Redox reactions can involve

   a. ions only. **c.** uncharged atoms only.

   b. molecules only. **d.** ions, molecules, or uncharged atoms.

CHAPTER 20 STUDY GUIDE FOR CONTENT MASTERY

Section 20.1 *continued*

In your textbook, read about determining oxidation numbers.

For each redox reaction below, determine the oxidation number of each element present. Write your answer above each symbol for the element.

9. $Cd(s) + NiO(s) \rightarrow CdO(s) + Ni(s)$

10. $Fe(s) + CuSO_4(aq) \rightarrow FeSO_4(aq) + Cu(s)$

11. $2Sb(s) + 3I_2(g) \rightarrow 2SbI_3(s)$

12. $2Cu_2S(s) + 3O_2(g) \rightarrow 2Cu_2O(s) + 2SO_2(g)$

13. $PbO_2(s) + Pb(s) + 2H_2SO_4(aq) \rightarrow 2PbSO_4(aq) + 2H_2O(l)$

14. $NH_4NO_3(s) \rightarrow 2H_2O(g) + N_2O(g)$

15. $Fe_2O_3(s) + 3CO(g) \rightarrow 2Fe(s) + 3CO_2(g)$

In your textbook, read about oxidation, reduction, oxidizing agents, and reducing agents.

Use your answers from questions 9–15 to fill in the following table for the listed reactions. For each reaction, show what is oxidized, what is reduced, the oxidizing agent, and the reducing agent.

Equation	Oxidized	Reduced	Oxidizing Agent	Reducing Agent
16. $Cd(s) + NiO(s) \rightarrow CdO(s) + Ni(s)$				
17. $Fe(s) + CuSO_4(aq) \rightarrow$ $FeSO_4(aq) + Cu(s)$				
18. $2Sb(s) + 3I_2(g) \rightarrow 2SbI_3(s)$				
19. $2Cu_2S(s) + 3O_2(g) \rightarrow$ $2Cu_2O(s) + 2SO_2(g)$				
20. $PbO_2(s) + Pb(s) + 2H_2SO_4(aq) \rightarrow$ $2PbSO_4(aq) + 2H_2O(l)$				
21. $NH_4NO_3(s) \rightarrow 2H_2O(g) + N_2O(g)$				
22. $Fe_2O_3(s) + 3CO(g) \rightarrow$ $2Fe(s) + 3CO_2(g)$				

CHAPTER 20   **STUDY GUIDE FOR CONTENT MASTERY**

Section 20.2 Balancing Redox Equations

In your textbook, read about balancing equations by using the oxidation number method.

Answer the following questions.

1. Why couldn't the oxidation number method be used for balancing the following equation?

   $KI(aq) + Pb(NO_3)_2(aq) \rightarrow PbI_2(s) + KNO_3(aq)$

   _____

   _____

2. The conventional method of balancing equations can be used to balance redox equations also. Why is it easier to use the oxidation number method to balance redox equations such as $Zn(s) + MnO_2(s) + H_2O(l) \rightarrow Zn(OH)_2(s) + Mn_2O_3(s)$?

   _____

   _____

   _____

   _____

3. Why might you sometimes use a combination of the oxidation number method and the conventional method to balance a redox equation?

   _____

   _____

   _____

4. If you are balancing an ionic redox equation, why is it important to know whether the reaction takes place in an acidic solution? How might your answer change if the reaction takes place in a basic solution?

   _____

   _____

   _____

   _____

5. What would be the advantage of using a net ionic equation to represent a redox reaction?

   _____

   _____

Write the numbers *1* through *5* to place in order the steps used to balance an equation by the oxidation number method.

_____ **6.** Determine the oxidation number of each element shown in the equation.

_____ **7.** Draw a line connecting the atoms involved in oxidation and another line connecting the atoms involved in reduction.

_____ **8.** If necessary, use the conventional method of balancing equations to balance all atoms and charges.

_____ **9.** Use coefficients in front of formulas in the equation to balance the number of electrons transferred in the redox part of the reaction.

_____ **10.** Write a chemical equation, showing all reactants and products in the reaction.

In your textbook, read about balancing net ionic redox equations by the oxidation number method.

Balance the following equations, using the oxidation number method for the redox part of the equation. If you need to, use the conventional method to balance the rest of the equation. Show your work.

11. $N_2(g) + H_2(g) \rightarrow NH_3(g)$

12. $S_8(s) + Cu(s) \rightarrow Cu_2S(s)$

13. $NO_3^-(aq) + Zn(s) \rightarrow Zn(OH)_4^{2-}(aq) + NH_3(aq)$ in a basic solution

CHAPTER 20 **STUDY GUIDE FOR CONTENT MASTERY**

Section 20.3 Half-Reactions

In your textbook, read about half-reactions.

In the space at the left, write *true* if the statement is true; if the statement is false, change the italicized word or phrase to make it true.

_____ **1.** A *species* is any kind of chemical unit involved in a process.

_____ **2.** Glucose and sucrose are different types of sugars. A solution of glucose, sucrose, and water contains exactly *two* different species.

_____ **3.** A half-reaction is part of a *decomposition* reaction.

_____ **4.** When magnesium reacts with oxygen, $Mg \rightarrow Mg^{2+} + 2e^-$ is the *reduction* half of the reaction.

_____ **5.** A species that undergoes *oxidation* will donate electrons to any atom that accepts them.

_____ **6.** A species can be a molecule, an atom, or an *electron*.

_____ **7.** Balancing equations by half-reaction is based on the number of *atoms* transferred.

_____ **8.** Balancing half-reactions involves balancing *both atoms and charge*.

_____ **9.** In writing an equation in ionic form, ionic compounds are written as *molecules*.

_____ **10.** The half-reaction $SO_2 + H_2O + 2e^- \rightarrow SO_4^{2-} + 4H^+$ shows that the reaction takes place in *a basic* solution.

In your textbook, read about identifying half-reactions.

For each of the following reactions, write the oxidation and reduction half-reactions. Identify each half-reaction as either oxidation or reduction. Then list the spectator ions that are present in the reaction. If no spectator ions are present, write *none*.

11. $Ca(s) + Al(NO_3)_3(aq) \rightarrow Al(s) + Ca(NO_3)_2(aq)$

_____

12. $NO_2(g) + NaOH(aq) \rightarrow NaNO_2(aq) + NaNO_3(aq) + H_2O(l)$

_____

13. $HCl(aq) + KMnO_4(aq) \rightarrow KCl(aq) + MnCl_2(aq) + Cl_2(g) + H_2O(l)$

_____

14. $H_3PO_3(aq) + HNO_3(aq) \rightarrow H_3PO_4(aq) + NO(g) + H_2O(l)$

_____

Section 20.3 *continued*

In your textbook, read about balancing equations using half-reactions.

Use your answers from questions 11–13 to help you balance these equations. Show your work.

15. $Ca(s) + Al(NO_3)_3(aq) \rightarrow Al(s) + Ca(NO_3)_2(aq)$

16. $NO_2(g) + NaOH(aq) \rightarrow NaNO_2(aq) + NaNO_3(aq) + H_2O(l)$

17. $HCl(aq) + KMnO_4(aq) \rightarrow KCl(aq) + MnCl_2(aq) + Cl_2(g) + H_2O(l)$

Balance the following equations, assuming all reactions take place in an acidic solution. Remember that charge, as well as atoms, must be balanced. Show your work.

18. $NO_3^-(aq) + H_2S(g) \rightarrow S(s) + NO(g)$

19. $Cr_2O_7^{2-}(aq) + I^-(aq) \rightarrow Cr^{3+}(aq) + I_2(s)$

CHAPTER 21 · STUDY GUIDE FOR CONTENT MASTERY

Electrochemistry

Section 21.1 Voltaic Cells

In your textbook, read about redox in electrochemistry.

Use each of the terms below just once to complete the passage.

voltaic	electrochemical cell	electric current	salt bridge	galvanic

Oxidation and reduction reactions can occur in separate solutions, as long as there are two

connections between the solutions. One connection is a(n) **(1)**_____

through which ions can flow. The other connection is a metal wire through which electrons

can flow. The flow of ions or electrons is known as a(n) **(2)**_____. The

complete setup, called a(n) **(3)**_____, can convert chemical energy into

electrical energy or electrical energy into chemical energy. These cells are also known as

(4)_____ cells or **(5)**_____ cells.

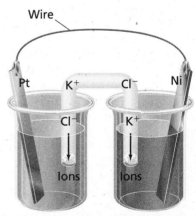

Use the diagram of an electrochemical cell to answer the following questions.

6. The equation at the bottom of each beaker shows the half-reaction that is occurring in that beaker. What kind of reaction (oxidation or reduction) is occurring in each beaker?

   Left beaker _____

   Right beaker _____

7. Write the net ionic equation for this electrochemical cell.

   _____

8. In which direction do electrons move through the wire?

   _____

9. What kind of ions (positive or negative) move from the ∩-shaped tube into each beaker?

   Left beaker _____   Right beaker _____

Section 21.1 *continued*

In your textbook, read about the chemistry of voltaic cells.

For each item in Column A, write the letter of the matching item in Column B.

Column A	Column B
_____ **10.** One of the two parts of an electrochemical cell, where either oxidation or reduction takes place	**a.** battery
_____ **11.** An electrode where oxidation takes place	**b.** electrical potential
_____ **12.** An electrode where reduction takes place	**c.** half-cell
_____ **13.** One or more electrochemical cells in a single package that generates electrical current	**d.** cathode
_____ **14.** A measure of the amount of current that can be generated from an electrochemical cell to do work	**e.** anode

In your textbook, read about calculating cell electrochemical potential.

Circle the letter of the choice that best completes the statement or answers the question.

15. The tendency of an electrode to gain electrons is called

   a. electron potential.   **c.** reduction potential.

   b. gravitational potential.   **d.** oxidation potential.

16. A sheet of platinum covered with finely divided platinum particles is immersed in a $1M$ HCl solution containing hydrogen gas at a pressure of 1 atm and a temperature of 25°C. The platinum sheet is known as a

   a. standard platinum electrode.   **c.** hydrogen chloride electrode.

   b. standard hydrogen electrode.   **d.** platinum chloride electrode.

17. The standard reduction potential of a half-cell is a measure of

   a. concentration.   **c.** temperature.

   b. pressure.   **d.** voltage.

18. Which of the following is the correct way to represent the equation, $H_2(g) + Cu^{2+}(aq) \rightarrow 2H^+(aq) + Cu(s)$?

   a. $H_2|H^+\|Cu^{2+}|Cu$ **c.** $Cu^{2+}|Cu\|H_2|H^+$

   b. $H^+|H_2\|Cu|Cu^{2+}$ **d.** $Cu|Cu^{2+}\|H^+|H_2$

19. When connected to a hydrogen electrode, an electrode with a negative standard reduction potential will carry out

   a. reduction.   **c.** both oxidation and reduction.

   b. oxidation.   **d.** neither oxidation nor reduction.

CHAPTER 21   **STUDY GUIDE FOR CONTENT MASTERY**

Section 21.1 *continued*

In your textbook, read about calculating cell electrochemical potential.

Use the table of standard reduction potentials below to answer the following questions.

Half-reaction	E^0 (volts)
$Al^{3+} + 3e^- \rightarrow Al$	−1.662
$Ga^{3+} + 3e^- \rightarrow Ga$	−0.549
$Tl^{3+} + 3e^- \rightarrow Tl$	+0.741

20. Suppose you have two voltaic cells whose half-cells are represented by the following pairs of reduction half-reactions. For each voltaic cell, identify which half-reaction will proceed in the forward direction as a reduction and which will proceed in the reverse direction as an oxidation.

Voltaic Cell #1

$Al^{3+}(aq) + 3e^- \rightarrow Al(s)$ _____

$Ga^{3+}(aq) + 3e^- \rightarrow Ga(s)$ _____

Voltaic Cell #2

$Tl^{3+}(aq) + 3e^- \rightarrow Tl(s)$ _____

$Ga^{3+}(aq) + 3e^- \rightarrow Ga(s)$ _____

21. Calculate the cell standard potential, E^0_{cell}, of each voltaic cell in question 20.

Voltaic Cell #1: _____

Voltaic Cell #2: _____

In your textbook, read about using standard reduction potentials.

Use the table of standard reduction potentials at the top of this page to answer the following questions.

22. Write the reduction and oxidation half-reactions for the following reaction:

$$Tl(s) + Al^{3+}(aq) \rightarrow Tl^{3+}(aq) + Al(s)$$

reduction half-reaction: _____

oxidation half-reaction: _____

23. What is the standard reduction potential, E^0, for each half-reaction in question 22?

$E^0_{reduction}$: _____ $E^0_{oxidation}$: _____

24. Calculate the cell standard potential, E^0_{cell}, for the reaction in question 22.

_____

25. Will the reaction in question 22 occur spontaneously as written? Explain why or why not.

_____

26. Will the reverse reaction, $Tl^{3+}(aq) + Al(s) \rightarrow Tl(s) + Al^{3+}(aq)$, occur spontaneously? Explain why or why not.

_____

Section 21.2 **Batteries**

In your textbook, read about dry cells, the lead–acid storage battery, and lithium batteries.

Complete the table below by writing the type of battery described on the right. Choose your answers from the following types: *lead–acid battery, lithium battery, mercury battery, nickel–cadmium battery, zinc–carbon dry cell.*

Type of Battery	Description
1.	Often used to power hearing aids and calculators because of its small size
2.	The standard, rechargeable automobile battery
3.	Often used in cordless drills, screwdrivers, and shavers because it is compact and rechargeable
4.	The most commonly used voltaic cell from the 1880s until recently
5.	Lightweight, long-lasting battery often used in watches and computers to maintain time and date settings

In the space at the left, write *true* **if the statement is true; if the statement is false, change the italicized word or phrase to make it true.**

_____ 6. In a zinc–carbon dry cell, a carbon rod serves as the *cathode*.

_____ 7. *Secondary* batteries produce electric energy by means of redox reactions that are not easily reversed.

_____ 8. One advantage of alkaline cells is that they are *larger* than dry cells.

_____ 9. When a lead–acid battery is generating electric current, *sulfuric* acid is consumed and lead(II) sulfate is produced.

_____ 10. Dry cells, alkaline cells, and mercury batteries are examples of *primary* batteries.

_____ 11. Compared to most other batteries, lithium batteries store a *small* amount of energy for their size.

_____ 12. *Secondary* batteries are rechargeable.

_____ 13. Each cell in a lead–acid battery generates about *12 volts*.

_____ 14. In a mercury battery, liquid mercury is *reduced*, forming mercury oxide.

_____ 15. Lead–acid batteries and nickel–cadmium batteries are examples of *secondary* batteries.

CHAPTER 21   **STUDY GUIDE FOR CONTENT MASTERY**

Section 21.2 *continued*

In your textbook, read about fuel cells.

Circle the letter of the choice that best completes the statement or answers the question.

16. The main purpose of a fuel cell is to produce

    a. fuel.   **b.** electric energy.   **c.** chemical energy.   **d.** heat.

17. In the hydrogen–oxygen fuel cell,

    a. hydrogen is oxidized and oxygen is reduced.   **c.** both oxygen and hydrogen are oxidized.

    b. oxygen is oxidized and hydrogen is reduced.   **d.** both oxygen and hydrogen are reduced.

18. What is the main difference between the reaction in a hydrogen–oxygen fuel cell and the burning of hydrogen in air?

    a. When hydrogen burns in air, the oxidation and reduction reactions are separated.

    b. The burning of hydrogen in air does not produce water.

    c. The reaction in a fuel cell does not produce water.

    d. The reaction in a fuel cell is very controlled.

In your textbook, read about corrosion.

Use the diagram below to answer the following questions.

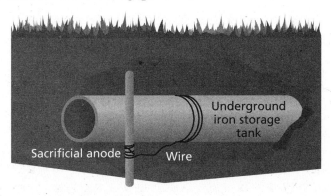

19. What is the function of the sacrificial anode?

_____

20. Name one metal that is commonly used as a sacrificial anode.

_____

21. Galvanizing the iron tank (or pipe) would serve the same function as a sacrificial anode. What is galvanizing?

_____

22. In what two ways does galvanizing protect iron?

_____

_____

CHAPTER 21 **STUDY GUIDE FOR CONTENT MASTERY**

Section 21.3 Electrolysis

In your textbook, read about reversing redox reactions and electrolysis.

In the space at the left, write the word or phrase in parentheses that correctly completes the statement.

_____ **1.** When a battery is being recharged, its redox reaction is reversed and energy is (absorbed, released) by the battery.

_____ **2.** The use of electrical energy to cause a chemical reaction is called (combustion, electrolysis).

_____ **3.** An electrochemical cell in which electrolysis is occurring is called an (electrolytic, exothermic) cell.

_____ **4.** In a Down's cell, sodium metal and chlorine gas are produced from (molten, solid) sodium chloride.

_____ **5.** The electrolysis of brine involves applying current to an aqueous solution of (hydrochloric acid, sodium chloride).

_____ **6.** The commercially important products of the electrolysis of brine are hydrogen gas, chlorine gas, and (oxygen gas, sodium hydroxide).

In your textbook, read about the purification of metallic ores, electroplating, and aluminum manufacture.

Answer the following questions.

7. Copper can be produced by heating Cu_2S in the presence of oxygen. Why must the copper then be subjected to electrolysis?

_____

_____

8. When an object is electroplated with silver, what is the anode and what is the cathode?

anode _____

cathode _____

9. The manufacture of aluminum begins with the electrolysis of aluminum oxide, Al_2O_3. What half-reaction occurs at the cathode?

_____

10. Why are plants that produce aluminum often built close to large hydroelectric power stations?

_____

_____

Name _____ Date _____ Class _____

Hydrocarbons

Section 22.1 Alkanes

In your textbook, read about organic chemistry, hydrocarbons, and straight-chain alkanes.

Use each of the terms below just once to complete the passage.

hydrocarbons	homologous series	organic compounds	straight-chain alkanes

   Most compounds that contain carbon are known as **(1)**_____. The

simplest group of such compounds are **(2)**_____, which contain only carbon

and hydrogen. If all of the carbon atoms are linked by single covalent bonds and there are no

branches, the compounds are called **(3)**_____. Ethane, propane, and

butane are three examples. They are members of one **(4)**_____ because they

differ from each other by a repeating unit ($— CH_2 —$).

In your textbook, read about branched-chain alkanes and naming them.

For each statement below, write *true* or *false*.

_____ **5.** The ability of carbon atoms to bond to two, three, or four other carbon
atoms makes possible a variety of branched-chain alkanes.

_____ **6.** A carbon atom or group of carbon atoms that branch off the main
hydrocarbon chain of an alkane is a substituent group.

_____ **7.** A skeletal formula is a way of representing an organic compound by
showing only the hydrogen atoms.

Use the IUPAC rules to name the following structures.

8. CH_3
   $CH_2CH_2CH_2CH_2CH_2CH_2CH_3$

9. CH_3
   $CH_3CHCH_2CH_2CH_2CHCH_3$
   CH_3
   CH_3

_____   _____

Draw the structure of each of the following alkanes.

10. 2-methylheptane

11. 2,3,4-trimethylpentane

Section 22.2 Cyclic Alkanes and Alkane Properties

In your textbook, read about cycloalkanes.

For each item in Column A, write the letter of the matching item in Column B.

Column A	Column B
_____ **1.** A simplified way of representing an organic compound by showing only the carbon-carbon bonds	**a.** *cyclo-*
_____ **2.** A way of representing an organic compound that saves space by not showing how the hydrogen atoms branch off the carbon atoms	**b.** condensed structural formula
_____ **3.** Indicates that a hydrocarbon has a ring structure	**c.** line structure
_____ **4.** A hydrocarbon that has a ring of carbon atoms in its structure	**d.** cyclic hydrocarbon

Use the IUPAC rules to name the following structure.

5. CH_3

         _____

      CH_2CH_3

Draw the structure of the following cycloalkane.

6. 1-ethyl, 2-propylcyclobutane

In your textbook, read about the properties of alkanes and multiple carbon-carbon bonds.

In the space at the left, write the word or phrase in parentheses that correctly completes the statement.

_____ **7.** All the bonds in an alkane are (polar, nonpolar).

_____ **8.** The attractive forces between alkane molecules are (stronger, weaker) than the attractive forces between alkane and water molecules.

_____ **9.** Alkanes are (very, not very) soluble in water.

_____ **10.** The boiling points of alkanes (increase, decrease) with increasing molecular mass.

_____ **11.** The chief chemical property of alkanes is their (low, high) reactivity.

_____ **12.** Alkanes are often used as (solvents, fuels) because they readily undergo combustion in oxygen.

_____ **13.** Alkanes are (saturated, unsaturated) hydrocarbons because they have only single bonds.

Section 22.3 Alkenes and Alkynes

In your textbook, read about alkenes, alkynes, and naming alkynes.

Use the following words to complete the statements.

alkene	alkyne	electron density	ethene	ethyne

1. An _____ is a hydrocarbon that has one or more triple covalent bonds between carbon atoms.

2. The unsaturated hydrocarbon _____ is the starting material for the synthesis of the plastic polyethylene.

3. An _____ is a hydrocarbon that has one or more double covalent bonds between carbon atoms.

4. Torches used in welding burn _____, which is commonly called acetylene.

5. Alkenes and alkynes are more reactive than alkanes because double and triple bonds have greater _____ than single bonds have.

Circle the letter of the correct name for each of the following structures.

6. $CH_3CH_2CH{=}CHCH_2CH_2CH{=}CH_2$
   a. 1,5-octadiene
   b. 3,7-octadiene
   c. 4,8-dioctene

7. $CH_3CH_2C{\equiv}CCH_2CH_3$
   a. 3-hexene
   b. 3-hexyne
   c. 3-pentyne

Use the IUPAC rules to name the following structures.

8. 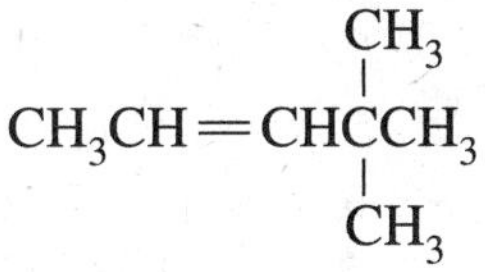
$$CH_3CH{=}CHCCH_3$$
with CH_3 above and CH_3 below the fourth carbon

9. 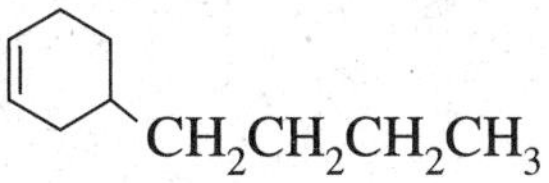 (cyclohexene ring) $CH_2CH_2CH_2CH_3$

_____

_____

10. 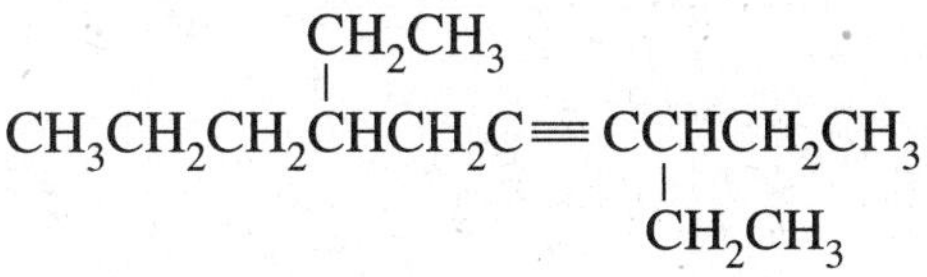
$$CH_3CH_2CH_2CHCH_2C{\equiv}CCHCH_2CH_3$$
with CH_2CH_3 on the fourth carbon and CH_2CH_3 below

11. 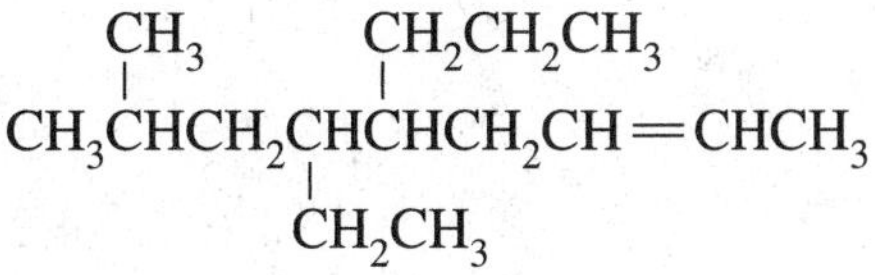
$$CH_3CHCH_2CHCHCH_2CH{=}CHCH_3$$
with CH_3, $CH_2CH_2CH_3$ above and CH_2CH_3 below

_____

_____

CHAPTER 22

STUDY GUIDE FOR CONTENT MASTERY

Section 22.4 **Isomers**

In your textbook, read about structural isomers, stereoisomers, chirality, and optical isomers.

Complete the concept map by writing the term below that fits the description at the right of each box.

chirality	geometric isomers	isomers	optical isomers
polarized light	stereoisomers	structural isomers	

1. _____ — same molecular formula but different molecular structures

include

2. _____ — atoms bonded in the same order but arranged differently in space

3. _____ — atoms bonded in different orders

include

4. _____ — four different groups arranged differently around the same carbon

5. _____ — different arrangements of groups around a double bond

have / *rotate the plane of*

6. _____ — right- or left-handedness of a molecule

7. _____ — waves all vibrate in the same plane

Identify the type of isomers represented by each of the following pairs of structures. Choose your answers from the following types: *geometric isomers, optical isomers, structural isomers*. (In item 8, the symbols W, X, Y, and Z represent hypothetical groups.)

8.
```
      X              Y
      |              |
  W — C — Y      W — C — X
      |              |
      Z              Z
```

_____

9.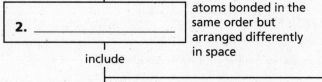

_____

10.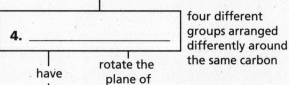

_____

CHAPTER 22

Section 22.5 Aromatic Hydrocarbons and Petroleum

In your textbook, read about the structure of benzene and aromatic compounds.

Use the following words to complete the statements.

aliphatic compounds	aromatic compounds	carcinogens

1. Substances that cause cancer are called _____.

2. Alkanes, alkenes, and alkynes are examples of _____.

3. All _____ contain benzene rings as part of their structure.

Circle the letter of the choice that best completes the statement or answers the question.

4. What is the molecular formula of benzene?

  a. C_6H_6 **b.** C_6H_{12} **c.** C_6H_{14} **d.** $C_{12}H_{12}$

5. Which of the following is the best way to represent the structure of benzene?

  a.   **b.**   **c.**   **d.**

6. Compared to alkenes and alkynes of similar size, benzene is

  a. less reactive.   **c.** slightly more reactive.

  b. about as reactive.   **d.** much more reactive.

7. The electrons in the ring of an aromatic compound are

  a. held tightly by one carbon nucleus.

  b. localized between specific carbon nuclei.

  c. shared equally by all of the carbon nuclei.

  d. shared by only three of the carbon nuclei.

8. The use of aromatic compounds should be limited because many of them

  a. produce chimney soot.   **c.** have pleasant smells.

  b. can cause health problems.   **d.** cannot be synthesized.

Use the IUPAC rules to name the following structures.

9. $CH_2CH_2CH_3$

10. CH_3

CH_2CH_3

_____   _____

CHAPTER 22   **STUDY GUIDE FOR CONTENT MASTERY**

Section 22.5 *continued*

In your textbook, read about natural sources of hydrocarbons and rating gasolines.

In the space at the left, write the word or phrase in parentheses that correctly completes the statement.

_____ **11.** (Petroleum, Natural gas) is a mixture of alkanes, aromatic hydrocarbons, and organic compounds containing sulfur or nitrogen atoms.

_____ **12.** The boiling of petroleum and collection of its components is called (sedimentation, fractional distillation).

_____ **13.** In the process known as (cracking, knocking), heavier petroleum fractions are converted to gasoline by breaking their large molecules into smaller ones.

_____ **14.** A gasoline's ability to burn evenly and prevent knocking is expressed by its (hexane, octane) rating.

Use the diagram of a fractionation tower to answer the following questions.

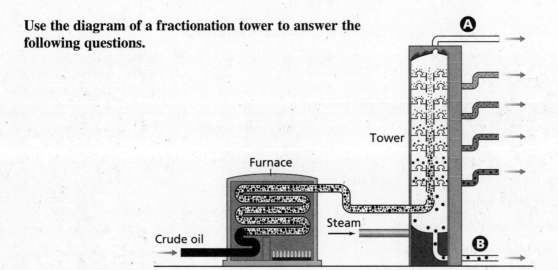

15. How does the temperature inside the tower vary from bottom to top?

_____

16. In what physical state is the material collected from pipe A? _____

17. What is the material collected from pipe A used for?

_____

18. What is the material collected from pipe B used for?

_____

19. Which pipe, A or B, collects hydrocarbons with higher boiling points? _____

20. Which pipe, A or B, collects smaller hydrocarbons? _____

CHAPTER 23 STUDY GUIDE FOR CONTENT MASTERY

Substituted Hydrocarbons and Their Reactions

Section 23.1 Functional Groups

In your textbook, read about functional groups.

Circle the letter of the choice that best completes the statement or answers the question.

1. In hydrocarbons, carbon atoms are generally linked to
   - **a.** other carbon atoms only.
   - **b.** hydrogen atoms only.
   - **c.** both carbon and hydrogen atoms.
   - **d.** atoms of any element.

2. Which of the following is an element commonly found in organic compounds?
   - **a.** nitrogen
   - **b.** argon
   - **c.** cesium
   - **d.** calcium

3. Atoms or groups of atoms, other than hydrogen and carbon, that occur in organic molecules and react in a certain way are called
   - **a.** functional groups.
   - **b.** polymers.
   - **c.** radicals.
   - **d.** monomers.

4. Which of the following is NOT a functional group?
   - **a.** a double bond
   - **b.** a triple bond
   - **c.** an alkane chain
   - **d.** a chlorine atom

In your textbook, read about organic compounds containing halogens.

Use each of the terms below just once to complete the passage.

alkyl halide	aryl halide	benzene	chiral	halocarbon	optical isomer

Any organic compound containing a fluorine, chlorine, bromine, or iodine substituent is

called a(n) **(5)** _____. Such a compound in which the substituent replaces a

hydrocarbon in a hydrocarbon containing only single bonds is called a(n)

(6) _____. If the substituent replaces a hydrogen bonded to an aromatic

compound such as **(7)** _____, the resulting compound is called a(n)

(8) _____. If an organic compound contains four different groups attached

to the same carbon atom, the compound is referred to as a(n) **(9)** _____. In

that case, the carbon atom is called a(n) **(10)** _____ carbon.

Section 23.2 **Alcohols, Amines, and Ethers**

In your textbook, read about the structure and properties of alcohols.

Circle the letter of the choice that best completes the statement or answers the question.

1. An alcohol is an organic compound in which a hydrogen atom of a hydrocarbon has been replaced by

   a. a hydroxyl group.   **b.** an oxygen atom.   **c.** an NH_2 group.   **d.** a COOH group.

2. Which of the following suffixes is used in naming alcohols?

   a. *-al*   **b.** *-oic*   **c.** *-ol*   **d.** *-ane*

3. The alcohol produced commercially in largest quantity is

   a. methanol.   **b.** isopropyl.   **c.** ether.   **d.** ethanol.

4. Alcohol molecules are generally

   a. nonpolar.   **b.** ionic.   **c.** very slightly polar.   **d.** moderately polar.

5. The alcohol produced by yeasts is

   a. methanol.   **b.** ethanol.   **c.** isopropanol.   **d.** cyclohexanol.

6. The simplest alcohol is

   a. methanol.   **b.** ethanol.   **c.** isopropanol.   **d.** butanol.

7. A poisonous alcohol used as a solvent for certain plastics and in the manufacture of insecticides is

   a. butanol.   **b.** ethanol.   **c.** cyclohexanol.   **d.** isopropanol.

8. Which of the following describes the solubility of ethanol in water?

   a. completely insoluble   **b.** slightly soluble   **c.** immiscible   **d.** completely miscible

9. What intermolecular attraction gives alcohols many of their physical properties?

   a. London forces   **b.** hydrogen bonds   **c.** ionic forces   **d.** dipole-dipole forces

10. Denatured alcohol is

    a. a mixture of two alcohols.

    b. ethanol to which noxious solvents have been added.

    c. ethanol that has been distilled.

    d. ethanol diluted with water.

11. How is ethanol generally removed from a water solution?

    a. filtration   **b.** distillation   **c.** adsorption   **d.** precipitation

12. The position of the functional group in an alcohol is indicated in its name by a

    a. letter at the end.   **c.** number and dash at the end.

    b. letter at the beginning.   **d.** number and dash at the beginning.

Section 23.2 *continued*

In your textbook, read about the structure and properties of ethers and amines.

For each statement below, write *true* or *false*.

_____ **13.** An ether contains an oxygen atom bonded to two carbon atoms.

_____ **14.** Ethers generally have much lower boiling points than alcohols of the same size.

_____ **15.** Ethers generally are more soluble in water than are alcohols.

_____ **16.** Ether molecules form hydrogen bonds with each other.

_____ **17.** Amines contain nitrogen bonded to carbon.

_____ **18.** More than one amino group can be present in an amine molecule.

_____ **19.** Amines are typically acids.

_____ **20.** Volatile amines tend to have pleasant odors.

In your textbook, read about naming alcohols, amines, and ethers.

Match each of the lettered structures (a–l) to the following names.

_____ **21.** 1-butanol

_____ **22.** 2-butanol

_____ **23.** 1,1-butanediol

_____ **24.** 2,2-butanediol

_____ **25.** 1,2-butanediol

_____ **26.** butyl ether

_____ **27.** butylpropyl ether

_____ **28.** 1-butylamine

_____ **29.** 2-butylamine

_____ **30.** 1,1-butyldiamine

_____ **31.** 2,2-butyldiamine

_____ **32.** 1,2-butyldiamine

Section 23.3 Carbonyl Compounds

In your textbook, read about aldehydes and ketones.

Circle the letter of the choice that best completes the statement or answers the question.

1. In a carbonyl group, an oxygen atom is

    a. single-bonded to a carbon atom.

    b. double-bonded to a carbon atom.

    c. bonded to a hydrogen atom.

    d. bonded to a carbon atom and another oxygen atom.

2. Which of the following makes a compound a ketone?

    a. a carboxyl group at the end   **c.** a carbonyl group at the end

    b. a carboxyl group between carbon atoms   **d.** a carbonyl group between carbon atoms

3. The formal names of aldehydes end with the suffix

    a. *-one.*   **b.** *-al.*   **c.** *-oic.*   **d.** *-ane.*

4. Which of the following correctly expresses the order of solubility, from greatest to least, of aldehydes, alcohols, and alkanes?

    a. aldehydes, then alkanes, then alcohols   **c.** alcohols, then aldehydes, then alkanes

    b. aldehydes, then alcohols, then alkanes   **d.** alcohols, then alkanes, then aldehydes

5. Which of the following is true of ketones in comparison to aldehydes?

    a. Neither is polar, and they are equally reactive.

    b. Ketones are polar, but aldehydes are not, and ketones are less reactive.

    c. Both are polar, and ketones are more reactive.

    d. Both are polar, and ketones are less reactive.

In your textbook, read about carboxylic acids, esters, and amides.

Use each of the terms below just once to complete the passage.

amide	carboxyl	carboxylic acids	ester	*-oic*	proteins

    The COOH group is called a(n) **(6)**_____. That group is found in the organic

compounds called **(7)**_____. The formal name of such compounds is formed by

adding the suffix **(8)**_____ to the corresponding alkane, followed by the word

acid. A(n) **(9)**_____ is a compound in which the acidic hydrogen of an organic

acid is replaced by a carbon atom or hydrocarbon chain. A(n) **(10)**_____ is a

compound in which the –OH group of an organic acid is replaced by a nitrogen atom bonded to

other atoms. The functional group of such compounds is found in **(11)**_____.

CHAPTER 23 **STUDY GUIDE FOR CONTENT MASTERY**

Section 23.4 Other Reactions of Organic Compounds

In your textbook, read about categories of organic reactions.

Match the descriptions of reactions in Column A with the types of reactions in Column B.

Column A	Column B
_____ **1.** An atom or a group of atoms is replaced by other atoms or groups.	**a.** elimination
_____ **2.** An atom or a group of atoms is replaced by F, Cl, Br, or I.	**b.** dehydration
_____ **3.** Additional bonds are formed between adjacent carbon atoms by the removal of atoms from the carbons.	**c.** condensation
	d. substitution
_____ **4.** H atoms are removed.	**e.** addition
_____ **5.** Atoms that form water are removed.	**f.** dehydrogenation
_____ **6.** Other atoms bond to carbons that are double or triple bonded.	**g.** halogenation
_____ **7.** Molecules join to form a larger molecule, with the loss of a small molecule.	

In your textbook, read about predicting the products of organic reactions.

Circle the letter of the choice that best answers the question.

8. What would be the products of a dehydration reaction in which ethanol was the reactant?

   a. ethyne and water   **b.** ethyne and hydrogen   **c.** ethene and water   **d.** ethene and hydrogen

9. What would be the product of a hydrogenation reaction in which propene and hydrogen were the reactants?

   a. propyne   **b.** propane   **c.** propanol   **d.** propanal

10. Suppose that ethene and chlorine react to form 1,2-dichloroethane. What type of reaction would that be?

   a. addition   **b.** elimination   **c.** substitution   **d.** condensation

11. What kind of reaction is represented by the following equation, which contains structural formulas?

   a. addition

   b. elimination

   c. substitution

   d. condensation

$$H-\underset{\underset{H}{|}}{\overset{\overset{F}{|}}{C}}-\underset{\underset{H}{|}}{\overset{\overset{F}{|}}{C}}-H \rightarrow H-\underset{}{\overset{\overset{H}{|}}{C}}=\underset{}{\overset{\overset{H}{|}}{C}}-H + F-F$$

12. What kind of reaction occurs when an unsaturated fat is converted to a saturated fat?

   a. dehydrogenation   **b.** hydrogenation   **c.** dehydration   **d.** hydration

Section 23.5 Polymers

In your textbook, read about monomers and polymers.

Use each of the terms below just once to complete the passage.

addition	catalyst	cellulose	condensation	celluloid
monomer	polymer	water	polymerization	

A large molecule consisting of many repeating structural units is called a(n) **(1)**_____.

A reaction in which such a compound is produced is called a(n) **(2)**_____ reaction.

Each of the unit molecules from which such a large molecule is made is called a(n)

(3)_____. The natural polymer **(4)**_____, which is found in

wood fiber, was treated with nitric acid to produce the first plastic, **(5)**_____.

A substance called a(n) **(6)**_____ is often required to make a polymerization

proceed at a reasonable rate. In a(n) **(7)**_____ polymerization, all the atoms present

in the monomers are present in the product. In a(n) **(8)**_____ polymerization, the

product is formed with the loss of a small by-product, usually **(9)**_____.

In your textbook, read about polymerization and the properties of polymers.

For each statement below, write *true* or *false*.

_____ **10.** The repeating group of atoms formed by the bonding of monomers is called the structural unit of the polymer.

_____ **11.** The number of structural units in a polymer chain is represented by the letter *n*.

_____ **12.** Nylon is made by means of an addition polymerization.

_____ **13.** Polymers do not differ greatly in their properties.

_____ **14.** Thermosetting plastics are generally more difficult to recycle than are thermoplastic polymers.

_____ **15.** Most of the plastic waste produced in the United States is recycled.

Draw the following structure.

16. The monomer that reacts to make the polymer shown below

$$-CH_2-CH\overbrace{-CH_2-CH-}^{}CH_2-CH-$$
$$\qquad\;\; |\qquad\qquad |\qquad\qquad\quad |$$
$$\qquad\;\; CH_3\qquad\;\; CH_3 \rfloor_n\qquad CH_3$$

CHAPTER 24 · STUDY GUIDE FOR CONTENT MASTERY

The Chemistry of Life

Section 24.1 Proteins

In your textbook, read about protein structure.

For each item in Column A, write the letter of the matching item in Column B.

Column A	Column B
_____ **1.** An organic molecule that has an amino functional group and a carboxyl functional group	**a.** denaturation
_____ **2.** A chain of two or more amino acids linked together	**b.** protein
_____ **3.** The process in which a protein's natural three-dimensional structure is disrupted	**c.** peptide bond
_____ **4.** An organic polymer made of many amino acids linked together in a specific way	**d.** amino acid
_____ **5.** The amide bond that joins two amino acids	**e.** peptide

Use the diagram below to answer the following questions.

6. Which part (A, B, C, D, or E) is the carboxyl group of an amino acid? _____

7. Which part (A, B, C, D, or E) is the amino group of an amino acid? _____

8. Which part (A, B, C, D, or E) is the side chain of an amino acid? _____

9. Which part (A, B, C, D, or E) is a peptide bond? _____

10. Which part (A, B, C, D, or E) includes the atoms that combine to form water? _____

11. What kind of chemical reaction does this diagram show? _____

12. What kind of organic molecule is formed in this reaction? _____

Name _____ Date _____ Class _____

Section 24.1 *continued*

In your textbook, read about the functions of proteins.

Complete the table below by writing the type of protein described on the left. Choose your answers from the following types: *enzyme, hormone, structural protein, transport protein.*

Functions of Proteins	
Description of function	**Type of protein**
13. Forming a structure that is vital to an organism	
14. Carrying smaller molecules throughout the body	
15. Carrying signals from one part of the body to another	
16. Catalyzing a reaction in an organism	

For each statement below, write *true* **or** *false*.

_____ **17.** The active site on an enzyme is a specific place where substrates bind.

_____ **18.** Collagen is a protein hormone found in skin, tendons, and bones.

_____ **19.** Most enzymes raise the activation energy of reactions.

_____ **20.** Insulin and chorionic gonadotropin are examples of structural proteins.

_____ **21.** The reactants in an enzyme-catalyzed reaction are known as substrates.

_____ **22.** Enzymes slow down chemical reactions.

_____ **23.** Hemoglobin is a protein that transports oxygen through the body.

_____ **24.** After the substrates bind to an enzyme, the enzyme's active site changes shape slightly.

_____ **25.** The human body is unable to use proteins as a source of energy.

_____ **26.** Insulin can be synthesized in the laboratory.

Answer the following questions.

27. Why must a molecule have a specific shape if it is to be a substrate of an enzyme?

_____

_____

28. How does being large help an enzyme carry out its function?

_____

_____

Name _____ Date _____ Class _____

Section 24.2 Carbohydrates

In your textbook, read about the different kinds of carbohydrates.

Use each of the terms below just once to complete the passage.

polysaccharide	carbohydrate	disaccharide	monosaccharide

A compound that contains multiple hydroxyl groups as well as an aldehyde or ketone

functional group is called a **(1)**_____. A simple sugar, or

(2)_____, is the simplest kind of carbohydrate. Two simple sugars can be

linked together by a condensation reaction to form a **(3)**_____, such as

sucrose. A large carbohydrate polymer made from 12 or more simple sugars is known as

a **(4)**_____.

Circle the letter of the choice that best completes the statement or answers the question.

5. The major function of carbohydrates in living organisms is as a source of
   a. nitrogen. **b.** hydrogen. **c.** information. **d.** energy.

6. How many carbon atoms do most common monosaccharides have?
   a. 1 **b.** 2 or 3 **c.** 5 or 6 **d.** 9 or 10

7. Most simple sugars are water soluble because they have several
   a. polar groups. **b.** nonpolar groups. **c.** hydrogen atoms. **d.** carbon atoms.

8. In aqueous solution, monosaccharides exist as
   a. polymers. **c.** cyclic structures.
   b. open-chain structures. **d.** both open-chain and cyclic structures.

9. When two monosaccharides bond to form a disaccharide, the new bond is
   a. a peptide bond. **b.** an ether bond. **c.** an alcohol bond. **d.** a carbonyl group.

10. Glucose and fructose link to form the disaccharide known as
    a. maltose. **b.** galactose. **c.** sucrose. **d.** lactose.

11. Disaccharides are too large to be absorbed into the bloodstream, so they must be broken
    down into
    a. monosaccharides. **b.** polysaccharides. **c.** carbon dioxide. **d.** atoms.

12. Complex carbohydrates are known as
    a. disaccharides. **b.** polysaccharides. **c.** monosaccharides. **d.** simple sugars.

13. Starch, cellulose, and glycogen are all made from monomers of
    a. amylase. **b.** sucrose. **c.** lactose. **d.** glucose.

Section 24.3 **Lipids**

In your textbook, read about the different kinds of lipids.

Use the following terms to complete the statements.

fatty acid	phospholipid	steroid	wax
lipid	saponification	triglyceride	

1. In a reaction called _____, sodium hydroxide is used to hydrolyze the ester bonds of a triglyceride.

2. Combining a fatty acid with a long-chain alcohol produces a _____.

3. A _____ is a large, nonpolar, biological molecule.

4. A lipid with a four-ring structure is known as a _____.

5. A _____ is a triglyceride in which one of the fatty acids is replaced by a polar phosphate group.

6. A long-chain carboxylic acid is known as a _____.

7. When three fatty acids are bonded to a glycerol backbone through ester bonds, a _____ is formed.

For each statement below, write *true* or *false*.

_____ **8.** All lipids are soluble in water.

_____ **9.** Lipids are an extremely efficient way to store energy in living organisms.

_____ **10.** Most fatty acids have an odd number of carbon atoms.

_____ **11.** Fatty acids that contain no double bonds are called unsaturated.

_____ **12.** Saturated fatty acids have higher melting points than unsaturated fatty acids.

_____ **13.** All lipids contain one or more fatty acid chains.

_____ **14.** A typical cell membrane is a single layer of phospholipids.

_____ **15.** Saponification is a process for making soap out of fats and oils.

_____ **16.** The lipid bilayer of the cell membrane acts as a barrier.

_____ **17.** The wax that coats plant leaves prevents water loss.

_____ **18.** Cholesterol and vitamin D are steroids.

CHAPTER 24   STUDY GUIDE FOR CONTENT MASTERY

Section 24.4 Nucleic Acids

In your textbook, read about the structure of nucleic acids and of DNA.

Use the diagram of DNA to answer the following questions.

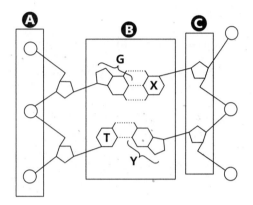

1. Which part (A, B, or C) represents the sugar molecules? _____

2. What is the name of these sugar molecules? _____

3. Which part (A, B, or C) represents the phosphate groups? _____

4. Which part (A, B, or C) represents the nitrogen bases? _____

5. What do the dotted lines represent? _____

6. If the part labeled G is guanine, what must the part labeled X be? _____

7. If the part labeled T is thymine, what must the part labeled Y be? _____

In your textbook, read about the function of DNA and about RNA.

Circle the letter of the choice that best completes the statement.

8. The main function of DNA is to store a cell's

  a. excess fat.   **c.** energy reserves.

  b. genetic information.   **d.** phosphate groups.

9. The instructions in a DNA molecule are carried in the form of a specific sequence of

  a. hydrogen bonds.   **b.** sugars.   **c.** phosphate groups.   **d.** nitrogen bases.

10. RNA is usually

  a. single stranded.   **b.** double stranded.   **c.** triple stranded.   **d.** a monomer.

11. The order of nitrogen bases in RNA determines the sequence of

  a. simple sugars in a polysaccharide.   **c.** amino acids in a protein.

  b. fatty acids in a triglyceride.   **d.** phosphate groups in DNA.

CHAPTER 24

STUDY GUIDE FOR CONTENT MASTERY

Section 24.5 Metabolism

In your textbook, read about anabolism and catabolism.

For each item in Column A, write the letter of the matching item in Column B.

Column A	Column B
_____ 1. The complete set of reactions carried out by an organism	**a.** catabolism
_____ 2. A nucleotide that functions as the universal usable energy form in living cells	**b.** metabolism
_____ 3. Metabolic reactions that synthesize complex molecules needed by an organism	**c.** anabolism
_____ 4. Metabolic reactions that break down complex biological molecules	**d.** ATP

For each statement below, write *anabolism* or *catabolism*.

_____ **5.** Starch is broken down into glucose monomers.

_____ **6.** Amino acids are linked by peptide bonds to form proteins.

_____ **7.** DNA is synthesized from free nucleotides.

_____ **8.** The proteins in food are broken down into individual amino acids.

_____ **9.** Three fatty acids combine with glycerol to make a triglyceride.

In your textbook, read about photosynthesis, cellular respiration, and fermentation.

For each statement below, write *true* or *false*.

_____ **10.** During photosynthesis, carbohydrates are made from oxygen and glucose.

_____ **11.** Plant cells can carry out photosynthesis, but animal cells cannot.

_____ **12.** The net equation for cellular respiration is the reverse of the net equation for photosynthesis.

_____ **13.** Fermentation is more efficient than cellular respiration at extracting energy from glucose.

_____ **14.** In alcoholic fermentation, glucose is converted to ethanol and carbon dioxide.

_____ **15.** Muscle cells carry out lactic acid fermentation when they have too much oxygen.

CHAPTER 25 **STUDY GUIDE FOR CONTENT MASTERY**

Nuclear Chemistry

Section 25.1 Nuclear Radiation

In your textbook, read about the terms used to describe nuclear changes.

Use each of the terms below just once to complete the passage.

alpha particle	radioactivity	gamma ray	radioisotope
beta particles	radiation	X ray	radioactive decay

The discovery of the **(1)**_____ in 1895 by Wilhelm Roentgen opened a whole new field of research. Among those who worked in this new field were Pierre and Marie Curie. The Curies discovered that some forms of matter give off **(2)**_____, a combination of particles and energy. Marie Curie named this process **(3)**_____. Another term used to describe the process by which one element spontaneously changes into another element is **(4)**_____. Any isotope that undergoes such changes is called a(n) **(5)**_____.

There are three common forms of radiation. One type is a form of energy known as **(6)**_____. The other types of radiation consist of particles. The form of radiation containing the heavier particle is made up of helium nuclei called **(7)**_____. The form of radiation containing the lighter particle consists of electrons called **(8)**_____.

In your textbook, read about the discovery of radioactivity.

Complete each statement.

9. Wilhelm Roentgen discovered the form of energy known as _____.

10. The form of nuclear radiation that has the greatest penetrating power is the _____.

11. When a radioactive nucleus gives off a gamma ray, its atomic number increases by _____.

12. The three types of radiation were first identified by _____.

13. Each alpha particle carries an electric charge of _____.

14. Each beta particle carries an electric charge of _____.

15. Each gamma ray carries an electric charge of _____.

CHAPTER 25 **STUDY GUIDE FOR CONTENT MASTERY**

Section 25.2 **Radioactive Decay**

In your textbook, read about the changes that take place in an atomic nucleus when it decays.

Circle the letter of the choice that best completes the statement.

1. The number of stable isotopes that exist compared to the number of unstable isotopes is

    a. much less.   **b.** much more.   **c.** slightly more.   **d.** about the same.

2. A lightweight isotope is likely to be stable if the ratio of protons to neutrons in its nucleus is

    a. 1:2.   **b.** 1:1.   **c.** 2:1.   **d.** 5:1.

3. The only nucleon among the following is the

    a. electron.   **b.** positron.   **c.** beta particle.   **d.** neutron.

4. The isotope least likely to be found in the band of stability among the following is

    a. $^{13}_{6}$C.   **b.** $^{17}_{8}$O.   **c.** $^{32}_{13}$Al.   **d.** $^{29}_{14}$Si.

5. The isotope formed by the beta decay of $^{40}_{19}$K has an atomic number of

    a. 18.   **b.** 39.   **c.** 20.   **d.** 21.

6. The isotope formed by the alpha decay of $^{238}_{92}$U has a mass number of

    a. 234.   **b.** 236.   **c.** 238.   **d.** 240.

7. The positron produced during positron emission comes from a(n)

    a. neutron.   **b.** proton.   **c.** electron.   **d.** positron.

8. During electron capture, a proton in the nucleus of an atom is converted into a(n)

    a. neutron.   **b.** positron.   **c.** electron.   **d.** another proton.

9. When the isotope $^{238}_{91}$Pa decays by beta emission, the isotope formed is

    a. $^{234}_{89}$Ac.   **b.** $^{238}_{90}$Th.   **c.** $^{237}_{92}$U.   **d.** $^{238}_{92}$U.

10. The isotope formed by the alpha decay of $^{154}_{66}$Dy is

    a. $^{150}_{66}$Dy.   **b.** $^{150}_{67}$Ho.   **c.** $^{150}_{64}$Gd.   **d.** $^{154}_{67}$Ho.

11. The neutron-to-proton ratio for the isotope sodium-23 is

    a. 1 : 1.1.   **b.** 1.1 : 1.   **c.** 2.1 : 1.   **d.** 1 : 2.1.

12. The decay of $^{162}_{69}$Tm yields $^{162}_{68}$Er and

    a. $^{4}_{2}$He.   **b.** $^{0}_{-1}$e.   **c.** γ.   **d.** $^{0}_{+1}\beta$.

13. Atoms located above the band of stability on a graph of numbers of neutrons versus number of protons are usually unstable because they contain too many

    a. protons.   **b.** neutrons.   **c.** electrons.   **d.** nucleons.

CHAPTER 25   **STUDY GUIDE FOR CONTENT MASTERY**

Section 25.3 Transmutation

In your textbook, read about the process of induced nuclear transmutation.

For each statement below, write *true* or *false*.

_____ **1.** Transmutation is the conversion of an atom of one element to an atom of another element.

_____ **2.** All nuclear reactions involve some type of nuclear transmutation.

_____ **3.** Scientists induce transmutations by bombarding stable nuclei with high-energy alpha, beta, or gamma radiation.

_____ **4.** The first induced nuclear transmutation was carried out by Marie and Pierre Curie in 1897.

_____ **5.** Most induced transmutation reactions are produced in high-energy particle accelerators.

_____ **6.** Neptunium and plutonium were the first transuranium elements discovered.

_____ **7.** The nuclear formula for a neutron is *n*.

_____ **8.** The half-life of a radioisotope is the time it takes for that isotope to decay.

_____ **9.** A radioisotope that decays very rapidly has a short half-life.

_____ **10.** Radioisotopes with very long half-lives are seldom found in Earth's crust.

_____ **11.** Temperature is the only factor that affects the half-life of a radioisotope.

_____ **12.** Carbon dating is not used to measure the age of rocks because the half-life of carbon-14 is too short.

_____ **13.** After an organism dies, its ratio of carbon-14 to carbon-12 and carbon-13 increases.

_____ **14.** Scientists currently believe that all of the possible transuranium elements have been discovered.

_____ **15.** When an atom of $^{125}_{52}\text{Te}$ is bombarded with protons, the products are $^{125}_{53}\text{I}$ and neutrons.

_____ **16.** Mass number and atomic number are conserved in all nuclear reactions.

_____ **17.** The mass of a 25.0 g piece of $^{238}_{96}\text{Cm}$ (half-life: 2.4 hr) will be reduced to 3.1 g after 7.2 hr.

Section 25.4 **Fission and Fusion of Atomic Nuclei**

In your textbook, read about the process of by which electrical energy is produced in a nuclear power plant.

Use the following diagram to complete the passage.

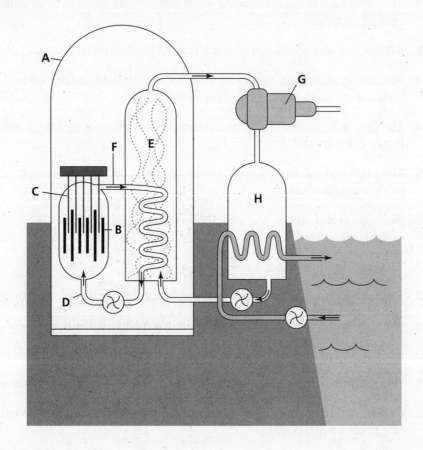

In a nuclear power plant, energy is produced in the reactor core by fission reactions

that occur in uranium-containing bars called **(1)**_____. The uranium is

found at location **(2)**_____ in the diagram. The rate at which the nuclear

reaction takes place is controlled by other bars called **(3)**_____. These bars

of metal are found at location **(4)**_____. One of the important safety factors

in the power plant is a strong dome-shaped structure surrounding the reactor. The structure is

labeled **(5)**_____ in this diagram and called **(6)**_____.

Section 25.4 *continued*

Heat produced by nuclear fission is carried away by **(7)**_____, which enters the core at point **(8)**_____ in the diagram. It then leaves the core at point **(9)**_____.

Heat from the reactor core is used to boil water in the **(10)**_____, shown at **(11)**_____ in the diagram. Steam produced here is used to generate electricity at point **(12)**_____ in the diagram. The steam is then cooled at location **(13)**_____ by water from an outside source.

For each statement, write *true* or *false*.

_____ **14.** A nuclear reactor produces energy from fuel rods containing uranium-238.

_____ **15.** The amount of energy produced for each kilogram of uranium is about the same as the amount of energy from a kilogram of coal.

_____ **16.** The only elements that can be used as fuel in a nuclear power plant are those in which a chain reaction can occur.

_____ **17.** If more than a critical mass is present in a sample, that sample is said to have supercritical mass.

_____ **18.** Water is the most common coolant used in a nuclear reactor.

_____ **19.** Nuclear power plants usually produce electricity.

_____ **20.** The purpose of the control rods in a nuclear reactor is to reflect neutrons back into the core.

_____ **21.** The production of energy in a nuclear reactor can be stopped by pulling out all control rods.

_____ **22.** A breeder reactor produces more fuel than it uses.

_____ **23.** The fission products produced in nuclear power plants are not radioactive.

_____ **24.** An uncontrolled chain reaction led to the nuclear accident in Chernobyl, Ukraine.

Section 25.5 Applications and Effects of Nuclear Reactions

In your textbook, read about the methods used to detect and measure radiation.

For each item in Column A, write the letter of the matching item in Column B.

Column A	Column B
_____ 1. Worn by workers to monitor radiation exposure	**a.** PET
_____ 2. Contains phosphors that detect radiation	**b.** Geiger counter
_____ 3. Radiation energetic enough to break apart atoms	**c.** 100–300 mrem
_____ 4. Uses a gas-filled metal tube to detect and measure radiation	**d.** ionizing radiation
_____ 5. A material that gives off light when struck by radiation	**e.** rad
	f. iodine-131
_____ 6. A method used to detect very small amounts of an element in a sample	**g.** rem
_____ 7. A radioisotope used to indicate the presence of an element in a sample	**h.** genetic damage
_____ 8. Used to detect disorders of the thyroid gland	**i.** neutron activation analysis
_____ 9. A procedure that uses positrons to detect many different medical disorders	**j.** scintillation counter
	k. radiotracer
_____ 10. This type of radiation easily penetrates human tissue.	**l.** gamma ray
_____ 11. Damage caused by radiation that affects a person, but not their offspring	**m.** somatic damage
_____ 12. Radiation damage that can affect chromosomes and offspring	**n.** phosphor
_____ 13. A unit used to measure the amount of radiation absorbed by a body	**o.** film badge
_____ 14. A unit used to measure the amount of damage done to a body	
_____ 15. The annual amount of radiation to which a person is normally exposed	

CHAPTER 26 STUDY GUIDE FOR CONTENT MASTERY

Chemistry in the Environment

Section 26.1 Earth's Atmosphere

In your textbook, read about the terms used to describe the physical and chemical properties of Earth's atmosphere.

Complete each statement.

1. In the troposphere, temperatures generally _____ with increasing altitude.

2. The region of the atmosphere closest to Earth's surface is the _____.

3. Acid rain can lead to the loss of _____ from soil.

4. When sunlight acts on pollutants in the air, _____ is formed.

5. Most of the liquid water in the troposphere is in the form of droplets found

   in _____.

6. When oxides of nitrogen or sulfur react with moisture in the air, _____ is formed.

7. The most harmful of the sulfur-containing compounds in the troposphere

   is _____.

8. The region of Earth's atmosphere below the mesosphere is the _____.

9. The colors of sunsets are due to _____ in the troposphere.

10. The thinning of the _____ layer was first observed in the mid-1980s over the South Pole.

11. Smog-producing pollutants enter the atmosphere when _____ are burned.

12. The envelope of gases that surrounds Earth is the _____.

13. In the stratosphere, temperatures generally _____ with increasing altitude.

14. High-energy solar radiation is absorbed during the processes of photoionization and

    _____ in the upper atmosphere.

15. The layer of Earth's atmosphere immediately above the mesosphere is the

    _____.

16. _____ radiation can break the bonds in DNA molecules.

Section 26.1 *continued*

In your textbook, read about the composition of the atmosphere and chemical changes that take place in it.

Circle the letter of the choice that best completes the statement or answers the question.

17. The top of the troposphere is at a height above Earth's surface of about

    a. 5 km.   **b.** 5 mi.   **c.** 15 km.   **d.** 15 mi.

18. The region of the atmosphere most distant from Earth's surface is the

    a. stratosphere.   **b.** thermosphere.   **c.** mesosphere.   **d.** exosphere.

19. The most abundant gas in Earth's atmosphere is

    a. hydrogen.   **b.** oxygen.   **c.** nitrogen.   **d.** carbon dioxide.

20. The layer of the atmosphere in which all of our weather occurs is the

    a. troposphere.   **b.** stratosphere.   **c.** mesosphere.   **d.** thermosphere.

21. What percentage by mass of all atmospheric gases is found in the troposphere?

    a. 25%   **b.** 50%   **c.** 75%   **d.** 99%

22. Which of the following gases is not a significant part of Earth's atmosphere?

    a. hydrogen   **b.** argon   **c.** oxygen   **d.** nitrogen

23. The only substance that exists as a solid, liquid, and gas in the atmosphere is

    a. nitrogen.   **b.** oxygen.   **c.** carbon dioxide.   **d.** water.

24. The product formed when solar radiation strikes an oxygen molecule is

    a. O_2.   **b.** O.   **c.** O^{2-}.   **d.** O_2^-.

25. One possible molecule formed by photoionization is

    a. N_2^+.   **b.** O.   **c.** N_2.   **d.** O_3.

26. Most of the oxygen about 150 km in the atmosphere exists in the form of

    a. O.   **b.** O_2.   **c.** O_3.   **d.** O^+.

27. The substance primarily responsible for absorbing high-energy ultraviolet radiation in the stratosphere is

    a. O_2.   **b.** N_2.   **c.** O.   **d.** O_3.

28. The chemicals thought to be primarily responsible for the loss of ozone from the stratosphere are

    a. CO_2 and CO.   **b.** NO and NO_2.   **c.** CFCs.   **d.** SO_3 and SO_2.

29. The pH range of unpolluted precipitation is about

    a. 6–7.6.   **b.** 6.9–7.1.   **c.** 5–9.   **d.** 13–14.

30. One of the major components of photochemical smog is

    a. O_3.   **b.** N_2.   **c.** O_2.   **d.** H_2.

CHAPTER 26

STUDY GUIDE FOR CONTENT MASTERY

Section 26.2 Earth's Water

In your textbook, read about the physical and chemical properties of the hydrosphere.

For each statement below, write *true* or *false*.

_____ 1. By far, the largest fraction of Earth's water is found in the oceans.

_____ 2. About 20 percent of Earth's water is available as liquid freshwater.

_____ 3. The energy that drives evaporation, condensation, and precipitation of water comes from nuclear reactions within Earth's crust.

_____ 4. The average salinity of ocean water is about 35 g dissolved salts per kg ocean water.

_____ 5. The most common ion found in seawater is Ca^{2+}.

_____ 6. Most of the dissolved salts in the oceans come from human activities.

_____ 7. Desalination as a process for purifying seawater can be used only where large amounts of solar energy are available.

_____ 8. Cooking and drinking are the primary uses of freshwater in American households.

_____ 9. Even though water taken from a running stream may look clear and clean, it may not be safe to drink.

_____ 10. Most of the pollution in freshwater sources comes from legal activities.

_____ 11. The two most common elements in polluted water are nitrogen and phosphorus.

_____ 12. The chemical most frequently used to kill bacteria in municipal water treatment plants is oxygen.

_____ 13. The term *hydrosphere* refers to water in Earth's atmosphere only.

_____ 14. The first step in purifying water involves the removal of heavy metals.

_____ 15. When chlorine is added to water, it forms hypochlorous acid (HClO).

_____ 16. Clouds form as water vapor evaporates from dust particles in the air.

_____ 17. Most liquid freshwater on Earth is found as surface water.

Section 26.3 **Earth's Crust**

In your textbook, read about the physical and chemical properties of the lithosphere.

Identify Earth's layers. Write the names of the layers on the lines provided.

1. _____

2. _____

3. _____

4. _____

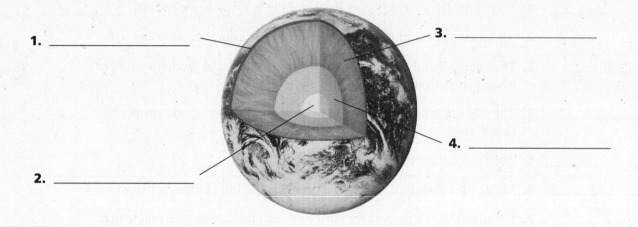

Complete each statement.

5. The probable age of Earth is about _____ years.

6. The thinnest part of Earth's structure is the _____.

7. _____ is the force that caused heavy elements to sink to the center of molten Earth.

8. The _____ is the solid part of Earth's crust.

9. Of all the elements in the newly-formed Earth's molten crust, the one that most readily escaped into outer space was _____.

10. The most abundant element in Earth's lithosphere is _____.

11. Most metallic elements found in Earth's crust are in the form of _____.

12. The most abundant element in Earth's core is _____.

13. A mineral that can be mined and purified economically is called a(n) _____.

14. The transition elements on the left side of the periodic table are most likely to occur in the lithosphere in the form of _____.

15. The most common form in which the alkaline earth metals occur in nature is as _____.

16. The layer of Earth that is most familiar to us because it is the only accessible layer is the _____.

CHAPTER 26

STUDY GUIDE FOR CONTENT MASTERY

Section 26.4 · Cycles in the Environment

In your textbook, read about the way elements are cycled through Earth's environment.

The diagram below shows how carbon is cycled through Earth's environment. Use the diagram and the information in Section 26.4 to answer the following questions.

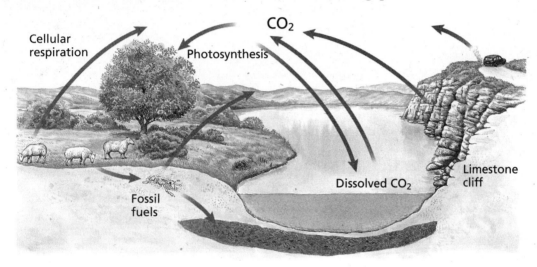

1. What form of carbon contributes to the greenhouse effect? _____

2. Name the process by which energy from the Sun is captured and stored in the bonds of carbon compounds. _____

3. Name the compounds in which plants store the Sun's energy. _____

4. By what process does carbon in the bodies of plants and animals return to Earth's crust?

   _____

5. In what form is carbon found deep within Earth's crust?

   _____

6. By what human activity are large quantities of carbon dioxide returned to the atmosphere?

   _____

7. What may the accumulation of large quantities of carbon dioxide lead to? _____

8. Name the process by which living plants and animals return carbon to the atmosphere.

   _____

9. What compound is the primary product of this process? _____

10. Name two ways in which carbon is stored in the hydrosphere.

   _____

Section 26.4 *continued*

The diagram below shows the steps in the nitrogen cycle. Use the diagram to answer the following questions.

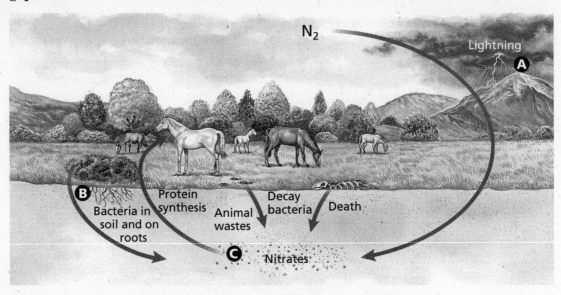

11. What is the name of the process that takes place at letter *A* in the diagram?

_____

12. What is the final product in this process?

_____

13. What is the name of the process indicated by letter *B* in the diagram?

_____

14. What is the final product in this process?

_____

15. What type of organism converts nitrogen gas into a form that plants can use?

_____

16. Where does this organism live?

_____

_____

17. How does nitrogen get from decaying plants and animals back into the atmosphere in the stage labeled *C* in the diagram?

_____

_____